Understanding and Using Rough Set Based Feature Selection: Concepts, Techniques and Applications

Muhammad Summair Raza • Usman Qamar

Understanding and Using Rough Set Based Feature Selection: Concepts, Techniques and Applications

Springer

Muhammad Summair Raza
Department of Computer Engineering,
College of Electrical & Mechanical Engineering
National University of Sciences and Technology (NUST)
Rawalpindi, Pakistan

Usman Qamar
Department of Computer Engineering,
College of Electrical & Mechanical Engineering
National University of Sciences and Technology (NUST)
Rawalpindi, Pakistan

ISBN 978-981-13-5278-2 ISBN 978-981-10-4965-1 (eBook)
DOI 10.1007/978-981-10-4965-1

Softcover reprint of the hardcover 1st edition 2017

Printed on acid-free paper

This Springer imprint is published by Springer Nature
The registered company is Springer Nature Singapore Pte Ltd.
The registered company address is: 152 Beach Road, #21-01/04 Gateway East, Singapore 189721, Singapore

Preface

Rough set theory (RST) has become a prominent tool for data science in various domains due to its analysis-friendly nature. From scientific discovery to business intelligence, both practitioners and scientists are using RST in various domains. Feature selection (FS) community is one to name. Various algorithms have been proposed in literature using RST, and a lot of search is still in progress.

For any practitioner and research community, this book provides a strong foundation of the concepts of RST and FS. It starts with the introduction of feature selection and rough set theory (along with the working examples) right to the advanced concepts. Sufficient explanation is provided for each concept so that the reader does not need any other source. A complete library of RST-based APIs implementing full RST functionality is also provided along with detailed explanation of each of the API.

The primary audience of this book is the research community using rough set theory (RST) to perform feature selection (FS) on large-scale datasets in various domains. However, any community interested in feature selection such as medical, banking, finance, etc. can also benefit from the book.

Rawalpindi, Pakistan — Muhammad Summair Raza
Rawalpindi, Pakistan — Usman Qamar

Contents

1 Introduction to Feature Selection 1
- 1.1 Feature 1
  - 1.1.1 Numerical 2
  - 1.1.2 Categorical Attributes 2
- 1.2 Feature Selection 3
  - 1.2.1 Supervised Feature Selection 4
  - 1.2.2 Unsupervised Feature Selection 6
- 1.3 Feature Selection Methods 8
  - 1.3.1 Filter Methods 8
  - 1.3.2 Wrapper Methods 10
  - 1.3.3 Embedded Methods 10
- 1.4 Objective of Feature Selection 11
- 1.5 Feature Selection Criteria 13
  - 1.5.1 Information Gain 13
  - 1.5.2 Distance 14
  - 1.5.3 Dependency 14
  - 1.5.4 Consistency 14
  - 1.5.5 Classification Accuracy 15
- 1.6 Feature Generation Schemes 15
  - 1.6.1 Forward Feature Generation 15
  - 1.6.2 Backward Feature Generation 16
  - 1.6.3 Random Feature Generation 17
- 1.7 Related Concepts 18
  - 1.7.1 Search Organization 18
  - 1.7.2 Generation of a Feature Selection Algorithm 18
  - 1.7.3 Feature Relevance 19
  - 1.7.4 Feature Redundancy 19
  - 1.7.5 Applications of Feature Selection 20
  - 1.7.6 Feature Selection: Issues 21

1.8 Summary 22
Bibliography 23

2 Background 27
2.1 Curse of Dimensionality 27
2.2 Transformation-Based Reduction 28
2.2.1 Linear Methods 29
2.2.2 Nonlinear Methods 33
2.3 Selection-Based Reduction 36
2.3.1 Feature Selection in Supervised Learning 36
2.3.2 Filter Techniques 37
2.3.3 Wrapper Techniques 39
2.3.4 Feature Selection in Unsupervised Learning 40
2.4 Correlation-Based Feature Selection 42
2.4.1 Correlation-Based Measures 43
2.4.2 Correlation-Based Filter Approach (FCBF) 44
2.4.3 Efficient Feature Selection Based on Correlation Measure (ECMBF) 46
2.5 Mutual Information-Based Feature Selection 46
2.5.1 A Mutual Information-Based Feature Selection Method (MIFS-ND) 48
2.5.2 Multi-objective Artificial Bee Colony (MOABC) Approach 48
2.6 Summary 50
Bibliography 50

3 Rough Set Theory 53
3.1 Classical Set Theory 53
3.1.1 Sets 53
3.1.2 Subsets 54
3.1.3 Power Sets 54
3.1.4 Operators 55
3.1.5 Mathematical Symbols for Set Theory 56
3.2 Knowledge Representation and Vagueness 56
3.3 Rough Set Theory (RST) 58
3.3.1 Information Systems 58
3.3.2 Decision Systems 59
3.3.3 Indiscernibility 59
3.3.4 Approximations 60
3.3.5 Positive Region 61
3.3.6 Discernibility Matrix 62
3.3.7 Discernibility Function 63
3.3.8 Decision-Relative Discernibility Matrix 63
3.3.9 Dependency 66
3.3.10 Reducts and Core 68

3.4 Discretization Process 70
3.5 Miscellaneous Concepts 72
3.6 Applications of RST 73
3.7 Summary 73
Bibliography 75

4 Advance Concepts in RST 81
4.1 Fuzzy Set Theory 81
4.1.1 Fuzzy Set 81
4.1.2 Fuzzy Sets and Partial Truth 83
4.1.3 Membership Function 83
4.1.4 Fuzzy Operators 84
4.1.5 Fuzzy Set Representation 86
4.1.6 Fuzzy Rules 86
4.2 Fuzzy-Rough Set Hybridization 88
4.2.1 Supervised Learning and Information Retrieval 89
4.2.2 Feature Selection 89
4.2.3 Rough Fuzzy Set 89
4.2.4 Fuzzy-Rough Set 91
4.3 Dependency Classes 92
4.3.1 Incremental Dependency Classes (IDC) 92
4.3.2 Direct Dependency Classes (DDC) 97
4.4 Redefined Approximations 102
4.4.1 Redefined Lower Approximation 102
4.4.2 Redefined Upper Approximation 104
4.5 Summary 106
Bibliography 106

5 Rough Set-Based Feature Selection Techniques 109
5.1 QuickReduct 109
5.2 Hybrid Feature Selection Algorithm Based on Particle Swarm Optimization (PSO) 112
5.3 Genetic Algorithm 113
5.4 Incremental Feature Selection Algorithm (IFSA) 115
5.5 Feature Selection Method Using Fish Swarm Algorithm (FSA) . . 116
5.5.1 Representation of Position 117
5.5.2 Distance and Centre of Fish 118
5.5.3 Position Update Strategies 119
5.5.4 Fitness Function 119
5.5.5 Halting Condition 119
5.6 Feature Selection Method Based on QuickReduct and Improved Harmony Search Algorithm (RS-IHS-QR) 119

5.7 A Hybrid Feature Selection Approach Based on Heuristic and Exhaustive Algorithms Using Rough set Theory (FSHEA) 120
5.7.1 Feature Selection Preprocessor 120
5.7.2 Using Relative Dependency Algorithm to Optimize the Selected Features 122
5.8 A Rough Set-Based Feature Selection Approach Using Random Feature Vectors 123
5.9 Summary 127
Bibliography 129

6 Unsupervised Feature Selection Using RST 131
6.1 Unsupervised QuickReduct Algorithm (USQR) 131
6.2 Unsupervised Relative Reduct Algorithm 134
6.3 Unsupervised Fuzzy-Rough Feature Selection 136
6.4 Unsupervised PSO-Based Relative Reduct (US-PSO-RR) 137
6.5 Unsupervised PSO-Based Quick Reduct (US-PSO-QR) 140
6.6 Summary 142
Bibliography 143

7 Critical Analysis of Feature Selection Algorithms 145
7.1 Pros and Cons of Feature Selection Techniques 145
7.1.1 Filter Methods 145
7.1.2 Wrapper Methods 146
7.1.3 Embedded Methods 146
7.2 Comparison Framework 147
7.2.1 Percentage Decrease in Execution Time 147
7.2.2 Memory Usage 147
7.3 Critical Analysis of Various Feature Selection Algorithms 148
7.3.1 QuickReduct 148
7.3.2 Rough Set-Based Genetic Algorithm 149
7.3.3 PSO-QR 150
7.3.4 Incremental Feature Selection Algorithm (IFSA) 151
7.3.5 AFSA 151
7.3.6 Feature Selection Using Exhaustive and Heuristic Approach 152
7.3.7 Feature Selection Using Random Feature Vectors 153
7.4 Summary 153
Bibliography 153

8 RST Source Code 155
8.1 A Simple Tutorial 155
8.1.1 Variable Declaration 156
8.1.2 Array Declaration 156
8.1.3 Comments 156
8.1.4 If-Else Statement 157

8.1.5 Loops . . . 157
8.1.6 Functions . . . 157
8.1.7 LBound and UBound Functions . . . 158
8.2 How to Import the Source Code . . . 158
8.3 Calculating Dependency Using Positive Region . . . 163
8.3.1 Main Function . . . 163
8.3.2 Calculate DRR Function . . . 164
8.3.3 SetDClasses Method . . . 166
8.3.4 FindIndex Function . . . 167
8.3.5 ClrTCC Function . . . 168
8.3.6 AlreadyExists Method . . . 169
8.3.7 InsertObject Method . . . 170
8.3.8 MatchCClasses Function . . . 171
8.3.9 PosReg Function . . . 172
8.4 Calculating Dependency Using Incremental Dependency Classes . . . 173
8.4.1 Main Function . . . 173
8.4.2 CalculateDID Function . . . 173
8.4.3 Insert Method . . . 176
8.4.4 MatchChrom Method . . . 177
8.4.5 MatchDClass Method . . . 178
8.5 Lower Approximation Using Conventional Method . . . 178
8.5.1 Main Method . . . 178
8.5.2 CalculateLAObjects Method . . . 180
8.5.3 FindLAO Method . . . 181
8.5.4 SetDConcept Method . . . 182
8.6 Lower Approximation Using Redefined Preliminaries . . . 183
8.7 Upper Approximation Using Conventional Method . . . 186
8.8 Upper Approximation Using Redefined Preliminaries . . . 187
8.9 QuickReduct Algorithm . . . 189
8.9.1 Miscellaneous Methods . . . 191
8.9.2 Restore Method . . . 192
8.9.3 C_R Method . . . 193
8.10 Summary . . . 194

About the Editors

Muhammad Summair Raza has PhD specialization in software engineering from the National University of Sciences and Technology (NUST), Pakistan. He completed his MS from International Islamic University, Pakistan, in 2009. He is also associated with the Virtual University of Pakistan as assistant professor. He has published various papers in international-level journals and conferences. His research interests include feature selection, rough set theory, trend analysis, software architecture, software design and non-functional requirements.

Usman Qamar has over 15 years of experience in data engineering both in academia and industry. He has a master's degree in computer systems design from the University of Manchester Institute of Science and Technology (UMIST), UK. His MPhil and PhD in computer science are from the University of Manchester. Dr. Qamar's research expertise is in data and text mining, expert systems, knowledge discovery and feature selection. He has published extensively in these subject areas. His post-PhD work at the University of Manchester involved various data engineering projects which included hybrid mechanisms for statistical disclosure and customer profile analysis for shopping with the University of Ghent, Belgium. He is currently an assistant professor at the Department of Computer Engineering, National University of Sciences and Technology (NUST), Pakistan, and also heads the Knowledge and Data Science Research Centre (KDRC) at NUST.

Chapter 1
Introduction to Feature Selection

This is an era of information. However, the data is only valuable if it is efficiently processed and useful information is derived out of it. It is now common to find applications that require data with thousands of attributes. Problem with processing such datasets is that they require huge amount of resources. To overcome this issue, research community has come up with an effective tool called feature selection. Feature selection lets us select only relevant data that we can use on behalf of the entire dataset. In this chapter we will discuss necessary preliminaries of feature selection.

1.1 Feature

A feature is a characteristic or an attribute of an object where an object is an entity having physical existence. A feature is an individual measurable property of a phenomenon being observed [1]. For example, features of a person can be "height", "weight", "hair colour", etc. A feature or combination of features helps us perceive about a particular aspect of that object, e.g. the feature "height" of a person helps us visualize physical height of the person. Similarly the feature "maximum load" gives us the information about the maximum upper bound of the force that a glass can bear with. For a particular concept or model, the quality of features is important. The better the quality of the feature, the better the resulting model [2].

In a dataset, data is represented in the form of matrix of "m" rows and "n" columns. Each row represents a record or object whereas each column represents a feature. So, in a dataset, each object is represented in the form of collection of features. For example, the dataset in Table 1.1 represents two features of five people.

M.S. Raza, U. Qamar, *Understanding and Using Rough Set Based Feature Selection: Concepts, Techniques and Applications*,
DOI 10.1007/978-981-10-4965-1_1

Table 1.1 Sample dataset

Name	Height	Eye colour
John	5.7	Blue
Bill	6.1	Black
Tim	4.8	Black
David	4.5	Blue
Tom	5.1	Green

It should be noted that all the data may not be in the form of such an arranged structure of rows and columns, e.g. DNA and protein sequences, in which case you may need to extract the features using different feature extraction techniques.

Based on the nature (domain) of data and type of the values a feature may take, features can be classified into two types:

(a) Numerical
(b) Categorical

1.1.1 *Numerical*

Numerical features, as a simple definition, are those that take numbers as their values. For example, in the case of height, 5 ft and 7 in. is a numerical value (however, if we mention height as tall, medium and short, then height will not be a numerical feature). Based on the range of values, numerical features can be discrete or continuous. Discrete values can be finite or infinite. Finite features are the ones that take limited set of values, e.g. "total matches played" and "total scores made". Infinite features are the ones for which the value set is infinite, e.g. total number of coin flips that result in heads. Continuous values on the other hand belong to the domain of real numbers, e.g. height of a person, weight in grams, etc.

Numerical features can be further classified into two types:

Interval scaled: Interval-scaled features are the ones where the difference between two values is meaning. For example, the difference between 90° and 70° is the same as the difference between 70° and 50° of temperature.

Ratio scaled: Ratio-scaled features have all the properties of interval-scaled features plus defined ratios for data analysis. For example, for attribute age, we can say that someone who is 20 years old is twice as old as someone who is 10 years old.

1.1.2 *Categorical Attributes*

In categorical attributes, symbols (words) are used to represent domain values. For example, gender can be represented by two symbols "M" and "F" or "male" or

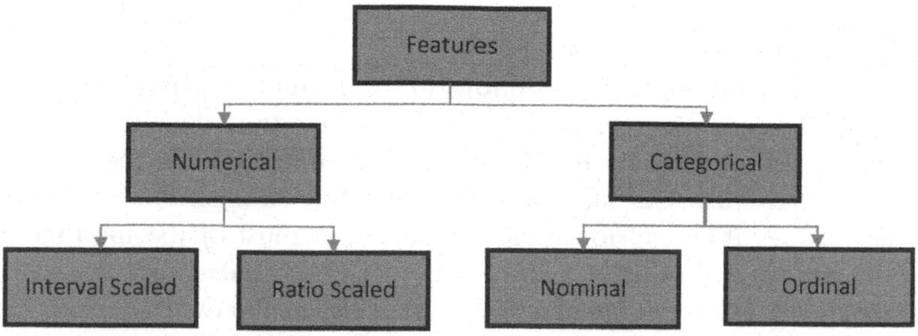

Fig. 1.1 Categorization of feature types

"female". Similarly qualification can be represented by "PhD", "BS", "MS", etc. Categorical attributes can be further classified into two types:

Nominal: Nominal features are the ones where order does not make sense. For example, the feature "gender" is a nominal feature because the domain values "M" and "F" do not involve any order. So comparison here only involves comparing domain values. Furthermore, only equal operator makes sense, i.e. we cannot compare values for less than or greater than operators.

Ordinal: In ordinal features, both equality and inequality can be involved, i.e. we can use "equal to", "less than" and "greater than" operators. For Example, "qualification" feature is an example of ordinal type as we can involve all of the above operators in comparison.

Figure 1.1 shows the categorization of the types of features.

1.2 Feature Selection

The amount of data to be processed and analyzed to get conclusions and to make decisions has significantly increased these days. Data creation is occurring at a record rate [3]. Data size is generally growing from day to day [4], and the rate at which new data are being generated is staggering [5]. This increase is seen in all areas of human activity right from the data generated on daily basis such as telephone calls, bank transactions, business tractions, etc., to more technical and complex data including astronomical data, genome data, molecular datasets, medical records, etc. It is the amount of information in the world which doubles every twenty (20) months [6]. These datasets may contain lot of information useful but still undiscovered.

This increase in data is twofold, i.e. both in number of samples/instances and number of features that are recorded and calculated. As a result many real-world applications need to process datasets with hundreds and thousands of attributes. This significant increase in the number of dimensions in datasets leads to a phenomenon called curse of dimensionality. The curse of dimensionality is the problem caused by the exponential increase in volume associated with adding extra dimensions to a (mathematical) space [7].

Feature selection is one of the solutions to the dilemma of curse of dimensionality. It is the process of selecting a subset of features from dataset that provides most of the useful information [6]. The selected set of features can then be used on behalf of the entire dataset. So, a good feature selection algorithm should opt to select the features that tend to provide complete or most of the information as present in the entire dataset and ignore the irrelevant and misleading features. Dimensionality reduction techniques can be categorized into two methods, "feature selection" and "feature extraction". Feature extraction techniques [8–20] project original feature space to a new feature space with lesser number of dimensions. The new feature space is normally constructed by combining the original feature some way. The problem with these approaches is that the underlying semantics of data are lost. Feature selection techniques [21–34] on the other hand tend to select features from the original features to represent the underlying concept. This lets feature selection techniques preserve data semantics. This book will focus on feature selection techniques.

Based on the nature of available data, feature selection can be categorized either as supervised feature selection or unsupervised feature selection. In supervised feature selection, the class labels are already provided and the feature selection algorithm selects the features on the basis of classification accuracy. In unsupervised feature selection, the class labels are missing and the feature selection algorithms have to select feature subset without label information. On the other hand, when class labels for some instances are given and missing for some, semi-supervised feature selection algorithms are used.

1.2.1 Supervised Feature Selection

The majority of real-world classification problems require supervised learning where the underlying class probabilities and class-conditional probabilities are unknown, and each instance is associated with a class label [35]. However, too often we come across the data having hundred and thousands of features. The presence of noisy, redundant and irrelevant features is another problem we have to commonly face in classification process. A relevant feature is neither irrelevant nor redundant to the target concept; an irrelevant feature is not directly associated with the target concept but affects the learning process, and a redundant feature does not add anything new to the target concept [35]. Normally it is difficult to filter out the important features especially when the data is huge. All this affects the performance and accuracy of the classifier. So, we have to preprocess data before feeding to classification algorithm. Supervised feature selection, at this point, plays its role to select minimum and relevant features. Figure 1.2 shows the general process of supervised feature selection.

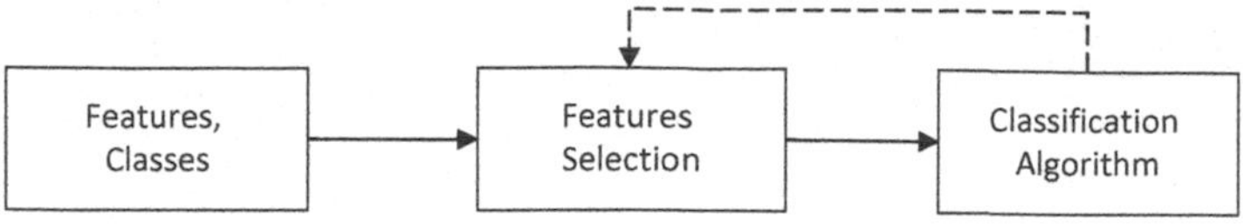

Fig. 1.2 Generic process of supervised feature selection

Table 1.2 A sample dataset in supervised learning

U	A	B	C	D	Z
X1	L	3	M	H	1
X2	M	1	H	M	1
X3	M	1	M	M	1
X4	H	3	M	M	2
X5	M	2	M	H	2
X6	L	2	H	L	2
X7	L	3	L	H	3
X8	L	3	L	L	3
X9	M	3	L	M	3
X10	L	2	H	H	2

As discussed earlier, in supervised feature selection, the class labels are already given along with features where each instance belongs to an already specified class label, and feature selection algorithm selects features from the original features based on some criteria. The selected features are then provided as input to the classification algorithm. The selection of features can or cannot be independent of the classification algorithm based on either it is filter-based or wrapper-based approach. This is realized by dotted arrow in above diagram from "classification algorithm" to "feature selection" process. Because the selected features are fed to classification algorithm, the quality of selected features affects the performance and accuracy of classification algorithm. The quality feature subset is the one that produces the same classification structure that is otherwise obtained by using entire feature set.

Table 1.2 shows a simple supervised dataset.

Here {A,B,C,D} are normal features and "Z" is a class label. We can perform feature selection by using many criteria, e.g. information gain, entropy, dependency, etc. If we use rough set-based dependency measure, then the dependency of "Z" on {A,B,C,D} is "100%", i.e. the feature set {A,B,C,D} uniquely determines the value of "Z". On the other hand, the dependency of "Z" on {A,B,C} is also "100%"; it means we can skip the feature "D" and use {A,B,C} only for further processing. A dependency simply determines how uniquely the value of an attribute "C" determines the value of an attribute "D". Calculation of rough set-based dependency measure is discussed in details in Chap. 3.

Table 1.3 A sample dataset in unsupervised learning

	X	Y	Z
A	1	2	1
B	2	3	1
C	3	2	3
D	1	1	2

1.2.2 Unsupervised Feature Selection

It is not necessary that classification information is given all the time. In unsupervised learning only the features are provided without any class label. The learning algorithm has to use only the available information. So a simple strategy may be to form the clusters (a cluster is a group of similar objects, similar to a classification structure except that class labels are not provided and hence only original features are used to form clusters, whereas in forming classification structures, class labels are given and used). Now again the problem of noisy, irrelevant and redundant data prohibits the use of all of the features to feed to learning algorithm; furthermore removing such features is again a cumbersome task for which manual filtration may not be possible. Again we have to take help from feature selection process. So common criteria is to select those features that give the same clustering structure as given by the entire set of features.

Consider the following example using hypothetical data. Table 1.3 shows the unsupervised dataset.

Dataset contains objects {A,B,C,D} where each object is characterized by features F = {X,Y,Z}.

Using k-means clustering, the objects {A,D} belong to cluster C_1 and objects {B, C} belong to cluster C_2. Now if computed, the feature subsets {X,Y}, {Y,Z} and {X,Z} can produce the same clustering structure. So, any of them can be used as the selected feature subset. It should be noted that we may have different feature subsets that fulfil the same criteria, so any of them can be selected, but efforts are made to find the optimal one, the one having minimum number of features.

The following is a description of how k-means clustering theorem works and how we calculated clusters.

K-Means Clustering Theorem

K-means clustering theorem is used to calculated clusters in unsupervised datasets. The steps of the theorem are given below:

Do {
Step 1: Calculate the centroid.
Step 2: Calculate the distance of each object from the centroids.
Step 3: Group objects based on minimum distance from the centroids.
} until no objects move from one group to another.

First of all we calculate the centroids; centroid is the centre point of a cluster. For the first iteration, we may assume any number of points as centroids. Then we calculate the distance of each point from the centroid; here distance is calculated using Euclidian distance measure. Once the distance of each object from each centroid is calculated, the objects are arranged in such a way that each object falls in the cluster whose centroid is close to that point. The iteration is repeated then by calculating new centroids of every cluster using all the points that fall in that cluster after an iteration is completed. The process continues until no point changes its cluster.

Now considering the datapoints given in the above datasets:

Step 1: We first suppose point A and B as two centroids C_1 and C_2.
Step 2: We calculate the Euclidean distance of each point from these centroids using the equation of Euclidean distance as follows:

$$d(xy) = \sqrt{(x_1 - x_2) + (y_1 - y_2)}$$

Note: x_1, y_1 are coordinates of point X and x_2, y_2 are coordinates of point Y. In our dataset the Euclidean distance of each point will be as follows:

$$D^1 = \begin{bmatrix} 0 & 1.4 & 2.8 & 1.4 \\ 1.4 & 0 & 2.4 & 2.4 \end{bmatrix} C_1 = \{1, 2, 1\} \text{ and } C_2 = \{2, 3, 1\}$$

Using "round" function,

$$D^1 = \begin{bmatrix} 0 & 1 & 3 & 1 \\ 1 & 0 & 2 & 2 \end{bmatrix} C_1 = \{1, 2, 1\} \text{ and } C_2 = \{2, 3, 1\}$$

In the above distance matrix, the value at row = 1 and column = 1 represents the distance of point A from the first centroid (here point A itself is the first centroid, so it is the distance of point A with itself which is zero), the value at row = 1 and column = 2 represents the Euclidean distance between point B and the first centroid, similarly the value at row = 2 and column = 1 shows Euclidean distance between point A and the second centroid, and so on.

Step 3: If we look at column 3, it shows that point "C" is close to the second centroid as compared to its distance from the first centroid. On the basis of the above distance matrix, the following groups are formed:

$$G^1 = \begin{bmatrix} 1 & 0 & 0 & 1 \\ 0 & 1 & 1 & 0 \end{bmatrix} C_1 = \{1, 2, 1\} \text{ and } C_2 = \{2, 3, 1\}$$

where the value "1" shows that point falls in that group, so from above group matrix, it is clear that points A and D are in one group (cluster), whereas points B and C are in another group (cluster).

Now the second iteration will be started. As we know that in the first cluster there are two points "A" and "D", so we will calculate their centroid as first step. So, $C1 = \frac{1+1}{2}, \frac{2+1}{2}, \frac{1+2}{2} = 1,2,2$. Similarly, $C_2 = \frac{2+3}{2}, \frac{3+2}{2}, \frac{1+3}{2} = 2,2,2$. On the basis of these new centroids, we calculate the distance matrix which is as follows:

$$D^2 = \begin{bmatrix} 1 & 2 & 2 & 1 \\ 1 & 1 & 1 & 1 \end{bmatrix} C_1 = \{1,2,2\} \text{ and } C_2 = \{2,2,2\}$$

$$G^2 = \begin{bmatrix} 1 & 0 & 0 & 1 \\ 0 & 1 & 1 & 0 \end{bmatrix} C_1 = \{1,2,2\} \text{ and } C_2 = \{2,2,2\}$$

Since $G^1 = G^2$, so we stop here. Note that the first column of D^2 matrix shows that point A is at equal distance from both clusters C_1 and C_2, so it can be placed in any of the cluster.

The above-mentioned clusters were formed considering all features {X,Y,Z}. Now if you perform the same steps using feature subsets {X,Y}, {Y,Z} or {X,Z}, the same clustering structure is obtained, which means that any of the above feature subset can be used by the learning algorithm.

1.3 Feature Selection Methods

Based on the relationship between selected features and learning algorithm or evaluation of selected feature subset, feature selection methods can be divided into three categories:

- Filter methods
- Wrapper methods
- Embedded methods

1.3.1 Filter Methods

Filter method is the most straight forward strategy for feature selection. In this method selection of features remains independent of learning algorithm, i.e. no feedback from the classification algorithm is used. Features are evaluated using some specific criteria using the intrinsic properties of the features. So the classification algorithm has no control over the quality of selected features, and thus poor quality may affect the performance and accuracy of the subsequent algorithm. They can be further classified into two types [36]:

1. Attribute evaluation methods
2. Subset evaluation methods

Fig. 1.3 Generic process of filter-based feature selection process

```
Input:
S - data sample with features X,|X| = n
J - evaluation measure to be maximized
GS – successor generation operator
Output:
Solution – (weighted) feature subset
L := Start_Point(X);
Solution := { best of L according to J };
repeat
L := Search_Strategy (L,GS(J),X);
X' := {best of L according to J };
if J(X')≥J(Solution) or (J(X')=J(Solution) and |X'| < |Solution|) then Solution :=X';
until Stop(J,L).
```

Fig. 1.4 Generic filter-based algorithm [37]

Attribute evaluation methods evaluate each individual feature according to the selected criteria and rank each feature after which a specific number of features are selected as the output. The subset evaluation methods on the other hand evaluate the complete subset on the basis of specified criteria. Figure 1.3 shows the generic diagram of filter-based approach [38].

You can note the unidirectional link between feature selection process and induction algorithm. Figure 1.4 shows the pseudocode of a generic feature selection algorithm.

We have a dataset having feature set X with n number of features. "J" is the measure that is maximized; it is basically the evaluation criteria on the basis of which features are selected, e.g. dependency, information gain, relevancy, etc. GS represents the successor generation operators which are used to generate the next feature subset from the current one. This operator may add or delete features from the current subset based on "J" measure. "Solution" is the subset that will contain the optimized feature subset output. Initially we assign "Solution" a starting feature subset. The starting subset may be empty (in case of forward feature selection strategy), may contain entire feature subset (in case of backward feature selection strategy) or random features (in case of random algorithms).

We get the next "L" features from the initially assigned feature subset, then optimize "L" into "X". Solution is then assigned "X" if "X" is more appropriate than the previous one. This step ensures that "Solution" always contains optimized feature subset and that we gradually refine the feature subset. The process continues until we meet the stopping criteria.

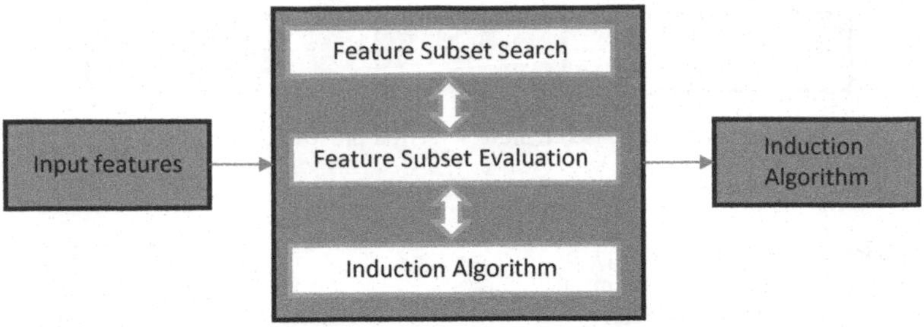

Fig. 1.5 Generic process of wrapper-based feature selection process

1.3.2 Wrapper Methods

Filter methods select optimal features independent of the classifier; however, optimal and quality feature subsets are dependent on the heuristics and biases of the classification algorithm, so it should be aligned with and selected on the basis of underlying classification algorithm. This is the underlying concept of wrapper approaches. So, in contrast with the filter methods, wrapper approaches do not select features independent of the classification algorithm. The feedback of the classification algorithm is used to measure the quality of selected features and thus results in high quality and performance of classifier. Figure 1.5 shows the generic model of wrapper approach [38].

It can be seen that there is a two-way link between "feature subset search", "evaluation" and "induction" processes, i.e. features are evaluated and thus searched again on the basis of the feedback from induction algorithm. From the figure, it is clear that wrapper approaches consist of three steps:

Step 1: Search feature subset.
Step 2: Evaluate features on the basis of induction algorithm.
Step 3: Continue process until we get optimized feature subset.

Clearly, induction algorithm works as a black box where the selected feature subset is sent and the acknowledgement is received in the form of some quality measure, e.g. error rate.

1.3.3 Embedded Methods

Embedded methods tend to overcome drawbacks of both filter and wrapper approaches. Filter methods evaluate features independent of classification algorithm, while wrapper-based approaches, using feedback from classifier, are computationally expensive as classifier is run many times to select optimal feature subset. Embedded models construct feature subsets as part of the classifier, so they

Table 1.4 Comparison of filter, wrapper and embedded approaches

Feature selection method	Advantages	Disadvantages
Filter	Simple approach	No interaction with classifier, so quality of selected features may affect quality of classifier
	Computationally less expensive	
Wrapper	Considers feature dependencies	Computationally expensive than filter approach
	Feature selection involves feedback from classifier thus resulting high-quality features	High overfitting probability
Embedded	Combines advantages of both filter and wrapper approach	Specific to learning machine

observe advantages of both wrapper approaches (feature subset evaluation is not independent of classifier) and filter method (efficient than wrapper approaches, furthermore selected features are evaluated using independent measures as well). A typical embedded method works as follows:

Step 1: Initialize feature subset (either empty or containing all the features).
Step 2: Evaluate the subset using independent measure.
Step 2.1: If it fulfils criteria more than current subset, this becomes the current subset.
Step 3: Evaluate the subset against evaluation criteria specified by classifier.
Step 3.1: If it fulfils criteria more than current subset, this becomes the current subset.
Steps 4: Repeat Step 2 to Step 3 until criteria are met.

There are three types of embedded methods [39]. The first are pruning methods that initially train the model using the entire set of features and then remove features gradually, then there are models that provide built-in mechanism to perform feature selection, and finally there are regularization models that minimize fitting errors and simultaneously remove the features by forcing the coefficients to be small or zero.

Table 1.4 shows the advantages and disadvantages of each approach.

1.4 Objective of Feature Selection

The main objective of feature selection is to select a subset of features from the entire dataset that could provide the same information, otherwise provided by the entire feature set. However different researchers describe feature selection from different perspectives. Some of these are:

Table 1.5 Symptoms table

Symptoms	Temperature	Flue	Cough	Sick
S1	H	Y	Y	Y
S2	H	Y	N	Y
S3	H	N	Y	Y
S4	N	Y	Y	Y
S5	N	N	N	N
S6	N	Y	N	Y
S7	N	N	Y	N

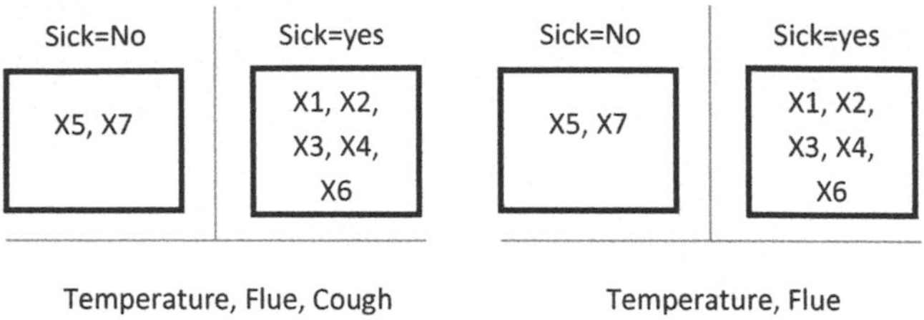

Fig. 1.6 Classification using the entire feature set vs selected features

Note that this was a very simple example, but in the real world, we come across the applications that require to process hundreds and thousands of the attributes before performing further tasks; feature selection is a mandatory preprocessor for all such applications.

1. Faster and more cost-effective models: Feature selection tends to provide minimum number of features to subsequent processes, so that these processes don't need to process the entire set of features, e.g. consider the classification with 100 attributes vs classification with ten attributes. The reduced number of features means the minimum execution time for the model, for example, consider the following dataset given in Table 1.5.

   Now if we perform classification on the basis of attributes temperature, flue and cough, we know that a patient having any of the symptoms S1, S2, S3, S4 and S6 will be classified as sick. On the other hand, the patient with symptoms S5 and S7 will not be sick. So to classify a patient as sick or not, we have to consider all the three attributes. However, note that if we take only the attributes "temperature" and "flue", we can also correctly classify the patient as sick or not; consequently to decide about the patient, the selected features "temperature" and "flue" give us the same information (accurately classify the records) as otherwise obtained by the entire set of attributes. Figure 1.6 shows the classification obtained in both cases, i.e. using all features and using selected features.
2. Avoid overfitting and improve performance: By selecting the best features that provide most of the information and by removing noisy, redundant and irrelevant features, the accuracy and effectiveness of model can be enhanced. It reduces the number of dimensions and thus enhances the performance of

subsequent model. It helps reduce noisy, redundant and irrelevant features thus improving data quality and enhancing accuracy of the model.

If we consider the dataset given in Table 1.5, it can be clearly seen that using two features, i.e. "temperature" and "flue", gives us the same result as given by the entire feature set, so ultimately the performance of the system will be enhanced. Furthermore, note that including or excluding the feature "cough" does not affect the classification accuracy and hence the final decision, so here it will be considered as redundant and thus can safely be removed.

3. Deeply understand the process that generated data: Feature selection also provides opportunity to understand the relationships between attributes to better understand the underlying process. It helps understand the relationship between the features and about the process that generated data. For example, in Table 1.5, it can be clearly seen that the decision class "sick" fully depends on the feature subset "temperature, flue". So, to accurately predict about a patient, we must have the accurate values of these two features.

   It should be noted that no other combination of the features (except the entire feature set, i.e. including all three features) gives us the exact information. If we consider the combination "temperature, cough", we cannot decide about a patient having temperature = N and cough = Y (S4 and S7 lead to different decisions for the same value of temperature = N and cough = Y). Similarly if we combine "flue, cough", we cannot decide about a patient having flue = N and cough = Y (S3 and S7 lead to different decision for the same value of flue = N and cough = Y). However, for any combination of "temperature, flue", we can precisely reach the decision. Hence there is deep relationship between the decision "sick" and the features "temperature, flue".

   It should also be noted that if an algorithm trained using the above dataset gets the input like temperature = N, flue = Y and sick = no, it should immediately conclude that the provided input contains erroneous data, which means that there was some mistake in mechanism collecting the data.

1.5 Feature Selection Criteria

The core of feature selection is the selection of good feature subset, but what actually makes a feature subset good? That is, what may be the criteria to select features from the entire dataset? Various criteria have been defined in literature for this purpose. We will discuss a few here.

1.5.1 Information Gain

Information gain [40] can be defined in terms of "uncertainty". The greater the uncertainty, the lesser the information gain. If IG(X) represents information gain from feature X, then feature X will be more good (hence preferred) if IG(X) > IG (Y). If "U" represents uncertainty function, $P(C_i)$ represents class C_i probability before considering feature "X" and $P(C_i|X)$ represents posterior probability of class C_i considering the feature "X", the information gain will be

$$IG(X) = \sum_i U(P(C_i)) - E\left[\sum_i U(P(C_i|X))\right]$$

That is, information gain can be defined as the difference of prior uncertainty and uncertainty after considering feature X.

1.5.2 Distance

Distance [40] measure specifies the discrimination power of feature, i.e. how strongly a feature can discriminate between classes. A feature with higher discrimination power is better than the one with less discrimination power. If C_i and C_i are two classes and X is the feature, then distance measure D(X) will be the difference of $P(X|C_i)$ and $P(X|C_j)$, i.e. difference of probability of "X" when class is C_i and when class is C_j. X will be preferred over Y if D(X) > D(Y). The larger the difference, the larger means the distance and more preferred the feature is. If $P(X|C_i) = P(X|C_i)$, feature X cannot separate classes C_i and C_j.

1.5.3 Dependency

Instead of the information gain or convergence power, dependency measure determines how strongly two features are connected with each other. In simple words dependency specifies about how uniquely the values of a feature determine values of other features. In case of supervised learning, it could be dependency of class label "C" on feature "X" while in case of unsupervised mode it may be dependency of other features on the one under consideration. We may select the features on which other features have high dependency value. If D(X) is the dependency of class C on feature X, then feature X will be preferred over feature Y if D(X) > D(Y).

1.5.4 Consistency

One of the criteria of feature selection may be to select those features that provide same class structure as provided by entire feature set. The mechanism is called consistency [40], i.e. to select the features with the condition $P(C|\text{Subset}) = P(C|$

Entire Feature Set). So we have to select the features that maintain same consistency as maintained by the entire dataset.

1.5.5 Classification Accuracy

Classification accuracy is normally dependent on the classifier used and is suitable for wrapper-based approaches. The intention is to select the features that provide the best classification accuracy based on the feedback from the classifier. Although this measure provides quality features (as classification algorithm is also involved), but it has few drawbacks, e.g. how to estimate accuracy and avoid overfitting, noise in the data may lead to wrong accuracy measure. It is computationally expensive to calculate accuracy as classifier takes time to learn from data.

1.6 Feature Generation Schemes

For any feature selection algorithm, the next feature subset generation is the key point, i.e. to select the members of feature subset for the next attempt if in the current attempt the selected feature subset does not provide appropriate solution. For this purpose we have four basic feature subset generation schemes.

1.6.1 Forward Feature Generation

In forward feature generation scheme, we start with an empty feature subset and add features one by one until feature subset meets the required criteria. The maximum size of feature subset selected may be equal to the total number of features in the entire dataset. A generic forward feature generation algorithm might look like the one given in Fig. 1.7.

"X" is the feature set comprising of all the features from dataset. There is a total of n number of features, so $|X| = n$; "Solution" is the final feature subset that will be selected by the algorithm.

In Step a, we have initialized "Solution" as an empty set; it does not contain any value. Step c adds the next feature in sequence to the "Solution". Step d evaluates the new feature subset (after adding the latest feature). If the condition is satisfied, the current feature subset is returned as selected subset; otherwise the next feature in the sequence is added to the "Solution". The process continues until the stopping criteria is met in which case the "Solution" is returned.

Input:
S - data sample with features X,|X| = n
J - evaluation measure
Initialization:
a) Solution ← {ϕ}
do
b) $\forall x \in X$
$c)\ Solution \leftarrow Solution \cup \{x_i\} \quad i = 1..n$
d) until Stop(J, Solution)
Output
d) Return Solution.

Fig. 1.7 A sample forward feature generation algorithm

1.6.2 Backward Feature Generation

Backward feature generation is opposite to forward generation scheme: in forward generation, we add with an empty set and add features one by one. In backward feature generation, on the other hand, we start with full feature set and keep on removing the features one by one, until no further feature can be removed without effecting the specified criteria. A generic backward feature generation algorithm is given in Fig. 1.8.

In backward feature generation algorithm, initially at Step a, the entire feature set is assigned to "Solution". Then we keep on removing features one by one. A new subset is generated after removing a feature. So, |Solution| ← |Solution| − 1. After removing the ith feature, the evaluation condition is checked. A typical condition might be that removing of a feature might not affect the classification accuracy of the remaining features or it might be that a feature can be removed if, after removing this feature, consistency of the remaining features remains the same as that of the entire feature set. Mathematically, if "R(Solution)" is consistent of a feature subset "Solution" and Xi is the ith feature in "Solution", then

If R(Solution – {Xi}) = R(Entire Feature Set) then
Solution = Solution – {Xi}

Note: We can also use hybrid feature generation approach in which case both forward generation scheme and backward generation scheme are followed concurrently. Two search procedures run in parallel where one search starts with empty set and the other starts with full set. One adds features one by one while the other removes features one by one. Procedure stops when any of the searches finds a feature subset according to specified condition.

Input:
S - data sample with features X,|X| = n
J - evaluation measure
Initialization:
a) Solution ← X
do
$b)$ $\forall x \in Solution$
$c)$ $Solution \leftarrow Solution - \{x_i\} \quad i = n..1$
d) until Stop(J, Solution)
Output
d) Return Solution.

Fig. 1.8 A sample backward feature generation algorithm

Input:
S - data sample with features X,|X| = n
J - evaluation measure

Solution ← $\{\phi\}$
For i=1 to n
If (Rand(0,1) < 0.5) then
Solution ← Solution $\cup \{X_i\}$
Until Stop(J, Solution)
Output
d) Return Solution.

Fig. 1.9 Sample hit-and-trial algorithm based on random feature generation algorithm

1.6.3 Random Feature Generation

There is another strategy called random feature generation. In random feature generation apart from all of the above-mentioned strategies, we randomly select features to be part of the solution. Features can be selected or skipped on any specified criteria. A simple procedure may be to select the features on the basis of random number generation mechanism. The following procedure generates a random number between 0 and 1; if the number is <0.5, the current feature will be included, or else it will be excluded.

$$Solution \leftarrow \{\phi\}$$
$$For\ i = 1\ to\ n$$
$$If(Rand(0,1) < 0.5)\ then$$
$$Solution \leftarrow Solution \cup \{X_i\}$$

On the basis of this mechanism, a typical hit-and-trial feature selection algorithm might look like the one given in Fig. 1.9.

Random feature generation approach is followed in genetic, particle swarm and fish swarm-like algorithms. However, in such algorithms some heuristics are also followed to generate next feature subset along with random generation mechanism. We will discuss it in the upcoming chapters.

1.7 Related Concepts

Now we will present some important preliminary concepts regarding feature selection.

1.7.1 Search Organization

For dataset d comprising feature X, a search algorithm needs to explore the feature space. The most intuitive method is to search the entire dataset and result any and find a candidate subset [6]. Unfortunately exhaustive search is not possible for datasets beyond smaller size and leads to expensive computational space. Hence exhaustive search is only suitable for datasets with smaller size.

An alternate way is to use random search [40] where a random set of features is selected and evaluated against their capability of being the required feature subset. The process continues until a potential candidate solution is found or a predefined time period has elapsed. Third and more commonly used methods are to imply heuristic-based search [40] for feature subset selection. In these methods a heuristic function is used to guide the search. The search is directed to achieve the maximum value of function. The process continues until we get a solution with required value of the heuristic function.

1.7.2 Generation of a Feature Selection Algorithm

Based on the above discussion, it is clear that feature selection algorithm should have the following three ingredients: feature subset generation component to generate next subset, evaluation measure to evaluate the quality of current subset and search organization to let the algorithm proceed on its way to perform feature selection.

Figure 1.10 shows a modified form of characteristic space of feature selection algorithm taken from [37].

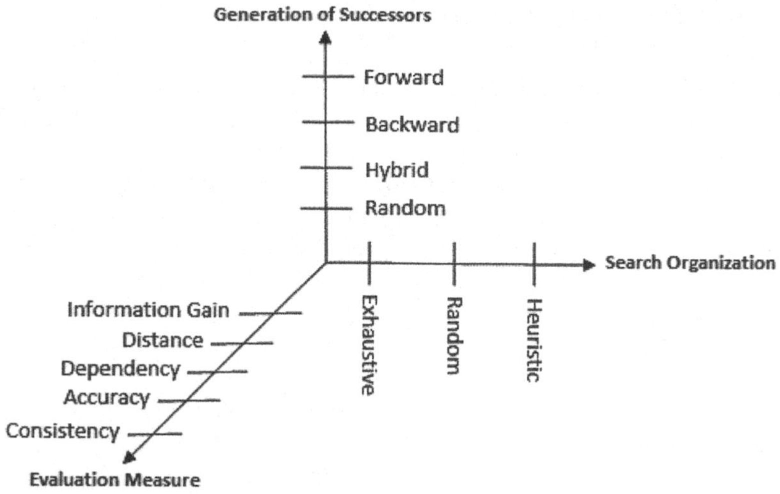

Fig. 1.10 Characteristics space of feature selection algorithm

1.7.3 Feature Relevance

Relevancy is defined in terms of the evaluation measure under consideration. A feature will be irrelevant if removing this feature does not affect the evaluation measure. For example, if D(*X*) is the dependency of class label on attribute set "X", then attribute $x_i \in X$ is irrelevant if $D(X - x_i) = D(X)$, i.e. removing the feature x_i does not affect the dependency of class label on the remaining feature set. Similarly it can be defined for other evaluation measures. Kohavi et al. [41] define the degree of relevance by stating three categories, strongly relevant, weakly relevant and irrelevant. Strongly relevant features are those which cannot be removed without distorting the underlying evaluation measure. Irrelevant features are those which can be safely removed without distorting the evaluation measure, and those that fall in between (e.g. features having dependency less than strongly relevant features and greater than zero, dependency of irrelevant features is considered to be zero) will be weakly relevant.

1.7.4 Feature Redundancy

Redundant features are those which do not add any information to feature subset. Just like irrelevant features, removal of which does not destroy the evaluation measure, redundant features are the ones adding which do not improve evaluation measure. If D(*X*) is the dependency of feature subset "X", then the feature Xj will be redundant if $D(X \cup \{x_j\}) = D(X)$. In other words if two features *X* and *Y* have the same dependency, then one of them is redundant.

1.7.5 Applications of Feature Selection

Today almost all fields of life are confronted by dilemma of high dimensionality. This is not the only problem, but the presence of noisy, irrelevant and redundant data makes it difficult to perform the intended analysis. In this situation, feature selection is an effective tool to deal with high-dimensional data to make it ready for further tasks. We will discuss here few domains where feature selection has played a major role in removing the above-mentioned problems.

1.7.5.1 Text Mining

With the passage of time and emergence of latest technologies day by day, we are overburdened with text, e.g. emails, blogs, books, etc. This requires proper processing of documents, e.g. to build vocabularies, to find conceptual depth of document regarding a particular concept (e.g. if a word "cricket" is repeated 15 times than any other word, then most probably the document explains something about "Cricket"), to group similar books, etc. Feature selection methods have been successfully applied both for text categorization and clustering. Authors in [44] provide five empirical evidences that feature selection methods can improve the efficiency and performance of text clustering algorithm. A number of algorithms have been proposed in literature to effectively use feature selection in text mining; [45–48] are few to name.

1.7.5.2 Image Processing

Representing images is not a straightforward task, as the number of possible image features is practically unlimited [49]. Be it simple surface image or domain of medical imaging, feature selection has played major role in dimensionality reduction. It can be used for processing/preprocessing of both images and video streams where it can help isolate a portion (features) of image for particular analysis purpose. Few applications of feature selection in medical image processing include image recognition, i.e. identification of important parts from the images in the presence of noise; image classification for retrieving images from large repositories on the basis of contents of images, for training purpose, e.g. to automatically distinguish between images of healthy and diseased optic nerves; and image cleansing to remove noise for further diagnostics.

1.7.5.3 Bioinformatics

The domain of bioinformatics is normally characterized by a large number of input features, e.g. there are millions of single nucleotide polymorphisms (SNPs), thousands of genes in microarrays, etc. So, an important task is to find out useful features (e.g. SNPs, genes, etc.) in various situations, e.g. to classify a certain community from the other or to diagnose a disease. Feature selection is an important tool that has successfully been deployed for this purpose. It helps focus on important or useful features only while ignoring the rest. It can be applied on microarray analysis, genomic analysis, mass spectrum analysis, etc.

1.7.5.4 Intrusion Detection

With the passage of time, the Internet has become a must-to-have utility for every person. Although getting benefits from it, one is always at risk of any type of intrusion while using the Internet. On daily basis intruders are coming with new ways to compromise your security. The static measures such as firewalls and antivirus applications are not necessary; we need to have more dynamic intrusion detection systems (IDSs); IDSs can be both host-based that monitor the local system resources, e.g. files, logs, disks, etc., and network-based systems that monitor network traffic. However, the problem with these systems is the huge amount of data passed through network. Here we can use feature selection process to detect important features of an event and then train the system on these features to automatically detect any intrusion in the future.

Apart from the above-mentioned domains, there are various other domains where feature selection has been successfully used, e.g. business and finance, industries, weather forecasting, remote sensing, network communication, etc. Covering all of these domains is out of scope of the book, so only few domains and use of feature selection in these domains have been discussed.

1.7.6 Feature Selection: Issues

Since its commencement, a lot of improvement has been made in feature selection process; however, there are various issues that are still challenging. We will discuss few issues here.

1.7.6.1 Scalability

The size of data is growing day by day; the increase is found both in the number of instances and number of features. It's common to have real-world applications with thousands of features and instances. The size has grown to the extent that it is difficult to fit the datasets in memory thus has posed a great challenge for feature selection algorithms to cater scalability of datasets. Majority of feature selection algorithms need the entire dataset to keep in memory for exact calculation which may demand more number of resources because it is hard for the algorithms to calculate ranks/relevance on the basis of less number of samples or alternatively calculate using different passes which may affect performance of these algorithms. So the conclusion is that with the rapidly growing size of datasets, it has become critical to address the scalability of feature selection algorithms as well.

1.7.6.2 Stability

Stability of feature selection algorithms is another challenge the community is focused on; feature selection algorithms may produce different features in case of small amount perturbation, whereas they are required to produce the same results, i.e. they should be stable. There are various factors that can affect the output of a specific feature selection algorithm and thus affect its sensitivity, e.g. dimensionality m, sample size n and different data distribution across different folds [39].

1.7.6.3 Linked Data

Almost all of the feature selection algorithms assume that data is independent and identically distributed; however, the important factor that is ignored is that data may be linked as well, especially with the emergence of social media like Facebook, Twitter, etc., where instances are linked with each other (users linked with posts, posts liked by other users, users make tweets, tweets are retweeted by other users, etc.). Unfortunately, the task of feature selection for linked data is rarely touched. The major challenges faced by feature selection algorithms in case of linked data are [39] the relations between the linked data and how to use these relations for feature selection. Although work has already been started in feature selection for linked data, e.g. [42, 43], however, it is still a challenge that feature selection community needs to be considered on priority basis.

1.8 Summary

In this chapter we have discussed the required preliminary concepts of feature selection. We started from the very basic concepts of features to different feature selection methods to underlying techniques to its applications and challenges. Normally the process of feature selection as explained in other books and tutorials comes with heavy mathematical equations, algebraic and statistical concepts which itself is a bigger challenge for a researcher especially the new comer in this domain to understand. Efforts are made to explain the concepts in very simple way and with examples where possible. We also provided the generic pseudocode of algorithms to give the idea of how a particular type of algorithm might work. This will be special help for the readers to understand more advance algorithm and develop their own.

Bibliography

1. Bishop CM (2006) Pattern recognition. Mach Learn 128:1–58
2. Domingos P (2012) A few useful things to know about machine learning. Commun ACM 55 (10):78–87
3. Villars RL, Olofson CW, Eastwood M (2011) Big data: what it is and why you should care. White Paper, IDC, p 14
4. Jothi N, Husain W (2015) Data mining in healthcare–a review. Proc Comput Sci 72:306–313
5. Kaisler S et al. (2013) Big data: issues and challenges moving forward. In: System sciences (HICSS), 2013 46th Hawaii international conference on, IEEE
6. Jensen R, Shen Q (2008) Computational intelligence and feature selection: rough and fuzzy approaches, vol 8. Wiley, Hoboken
7. Bellman R (1956) Dynamic programming and Lagrange multipliers. Proc Natl Acad Sci 42 (10):767–769
8. Neeman S (2008) Introduction to wavelets and principal components analysis. VDM Verlag Dr. Muller Aktiengesellschaft& Co. KG
9. Engelen S, Hubert M, Vanden Branden K (2016) A comparison of three procedures for robust PCA in high dimensions. Aust J Stat 34(2):117–126
10. Cunningham P (2008) "Dimension reduction." Machine learning techniques for multimedia. Springer, Berlin/Heidelberg, pp 91–112
11. Van Der Maaten L, Postma E, den Herik JV (2009) Dimensionality reduction: a comparative. J Mach Learn Res 10:66–71
12. Friedman JH, Stuetzle W (1981) Projection pursuit regression. J Am Stat Assoc 76 (376):817–823
13. Borg I, Groenen PJF (2005) Modern multidimensional scaling: theory and applications. Springer
14. Dalgaard P (2008) Introductory statistics with R. Springer
15. Zeng X, Luo S (2008) Generalized locally linear embedding based on local reconstruction similarity. Fuzzy systems and knowledge discovery, 2008. In: FSKD'08. Fifth international conference on, vol 5, IEEE
16. Saul LK et al (2006) Spectral methods for dimensionality reduction. Semisupervised Learning, 293–308
17. Liu R et al (2008) Semi-supervised learning by locally linear embedding in kernel space. Pattern recognition, 2008. ICPR 2008. 19th international conference on, IEEE
18. Gerber S, Tasdizen T, Whitaker R (2007) Robust non-linear dimensionality reduction using successive 1-dimensional Laplacian eigenmaps. In: Proceedings of the 24th international conference on machine learning, ACM
19. Donoho DL, Grimes C (2003) Hessian eigenmaps: locally linear embedding techniques for high-dimensional data. Proc Natl Acad Sci 100(10):5591–5596
20. Teng L et al (2005) Dimension reduction of microarray data based on local tangent space alignment. Cognitive informatics, 2005. (ICCI 2005). In: Fourth IEEE conference on, IEEE
21. Raman B, Ioerger TR (2002) Instance-based filter for feature selection. J Mach Learn Res 1 (3):1–23
22. Yan GH et al (2008) Unsupervised sequential forward dimensionality reduction based on fractal. In: Fuzzy systems and knowledge discovery, 2008. FSKD'08. Fifth international conference on, vol 2, IEEE

23. Tan F et al (2008) A genetic algorithm-based method for feature subset selection. Soft Comput 122:111–120
24. Loughrey J, Cunningham P (2005) Using early-stopping to avoid overfitting in wrapper-based feature selection employing stochastic search. In: Proceedings of the twenty-fifth SGAI international conference on innovative techniques and applications of artificial intelligence
25. Valko M, Marques NC, Castellani M (2005) Evolutionary feature selection for spiking neural network pattern classifiers. In: Artificial intelligence, 2005. epia 2005. portuguese conference on, IEEE
26. Huang J, Lv N, Li W (2006) A novel feature selection approach by hybrid genetic algorithm. PRICAI 2006: Trends in Artificial Intelligence, pp 721–729
27. Khushaba RN, Al-Ani A, Al-Jumaily A (2008) Differential evolution based feature subset selection. Pattern recognition, 2008. In: ICPR 2008. 19th international conference on, IEEE
28. Roy K, Bhattacharya P (2008) Improving features subset selection using genetic algorithms for iris recognition. In: IAPR Workshop on Artificial Neural Networks in Pattern Recognition. Springer, Berlin/Heidelberg
29. Dy JG, Brodley CE (2004) Feature selection for unsupervised learning. J Mach Learn Res:845–889
30. He X, Cai D, Niyogi P (2005) Laplacian score for feature selection. NIPS 186
31. Wolf L, Shashua A (2005) Feature selection for unsupervised and supervised inference: the emergence of sparsity in a weight-based approach. J Mach Learn Res 6(2): 1855–1887
32. Bryan K, Cunningham P, Bolshakova N (2005) Biclustering of expression data using simulated annealing. In: Computer-based medical systems, 2005. Proceedings. 18th IEEE symposium on, IEEE
33. Handl J, Knowles J, Kell DB (2005) Computational cluster validation in post-genomic data analysis. Bioinformatics 21(15):3201–3212
34. Gluck M (1985) Information, uncertainty and the utility of categories. In: Proceedings of the seventh annual conference on cognitive science society. Lawrence Erlbaum
35. Dash M, Liu H (1997) Feature selection for classification. Intellig Data Analy 1(1-4):131–156
36. Vanaja S, Ramesh Kumar K (2014) Analysis of feature selection algorithms on classification: a survey. Int J Comput Appl 96:17
37. Ladha L, Deepa T (2011) Feature selection methods and algorithms. Int J Comput Sci Eng 3 (5):1787–1797
38. John GH, Kohavi R, Pfleger K (1994) Irrelevant features and the subset selection problem. In: Machine learning: proceedings of the eleventh international conference
39. Tang J, Alelyani S, Liu H (2014) Feature selection for classification: a review. Data Classif Algor Appl 37
40. Hua J, Tembe WD, Dougherty ER (2009) Performance of feature-selection methods in the classification of high-dimension data. Pattern Recogn 42(3):409–424
41. Kohavi R, John GH (1997) Wrappers for feature subset selection. Artif Intell 97(1-2):273–324
42. Tang J, Liu H (2012) Feature selection with linked data in social media. In: Proceedings of the 2012 SIAM international conference on data mining. Society for Industrial and Applied Mathematics
43. Gu Q, Han J (2011) Towards feature selection in network. In: Proceedings of the 20th ACM international conference on information and knowledge management. ACM
44. Liu T et al (2003) An evaluation on feature selection for text clustering. ICML 3
45. Tutkan M, Ganiz MC, Akyokuş S (2016) Helmholtz principle based supervised and unsupervised feature selection methods for text mining. Inf Proc Manag 52(5):885–910
46. Özgür L, Güngör T (2016) Two-stage feature selection for text classification. Information sciences and systems 2015. Springer, pp 329–337
47. Liu M, Lu X, Song J (2016) A new feature selection method for text categorization of customer reviews. Commun Stat-Simul Comput 45(4):1397–1409

48. Kumar V, Minz S (2014) Multi-view ensemble learning for poem data classification using SentiWordNet. In: Advanced computing, networking and informatics, vol 1. Springer, pp 57–66
49. Bins J, Draper BA (2001) Feature selection from huge feature sets. In: Computer vision, 2001. ICCV 2001. Proceedings. Eighth IEEE international conference on, vol 2, IEEE

Chapter 2
Background

To overcome the phenomenon of curse of dimensionality, one of the methods is to reduce dimensions without effecting relevant information present in entire dataset. There are various techniques proposed in literature to reduce dimensions. In this chapter we have presented an overview of these techniques.

2.1 Curse of Dimensionality

Data creation is occurring at a record rate [1]. This increase is seen in all areas of human activity right from the data generated on daily basis such as telephone calls, bank transactions, business tractions, etc. to more technical and complex data including astronomical data, genome data, molecular datasets, medical records, etc. These datasets may contain lot of information useful but still undiscovered. This increase in data is twofold, i.e. both in number of samples/instances and number of features that are recorded and calculated. As a result many real world applications need to process datasets with hundreds and thousands of attributes. Few of such datasets are also publically available at [2].

The significant increase in the number of dimensions in datasets leads to the phenomenon called *curse of dimensionality*. The curse of dimensionality is the problem caused by the exponential increase in volume associated with adding extra dimensions to a (mathematical) space [3]. Dimension reduction (DR) is used as preprocessing [12]. The original feature space is mapped onto a new, reduced dimensionality space, and the samples are represented in that new space [4]. Normally datasets contain number of misleading and redundant information which needs to be removed before any further tasks can be performed on these datasets. For example, in case of deriving complex classification rules, it is more effective to first perform dimension reduction. This step not only enhances the performance but

M.S. Raza, U. Qamar, *Understanding and Using Rough Set Based Feature Selection: Concepts, Techniques and Applications*,
DOI 10.1007/978-981-10-4965-1_2

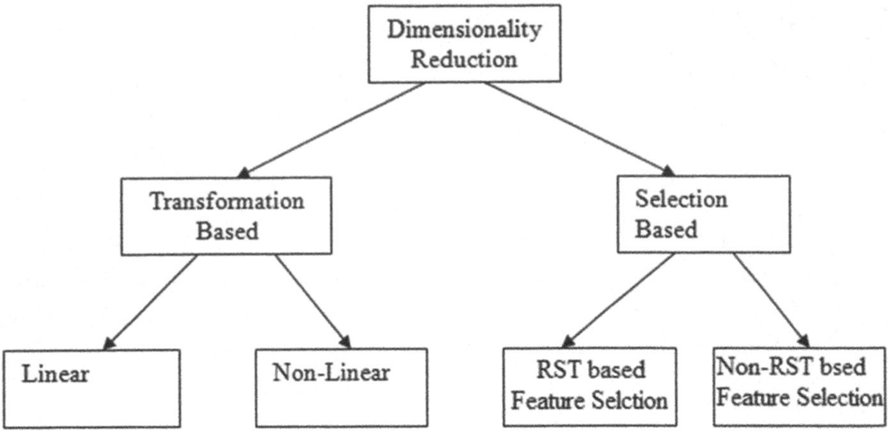

Fig. 2.1 Taxonomy of dimensionality reduction

also increases the resulting classification accuracy and making the rules more comprehensible.

There are various techniques to perform DR, e.g. [5–9], but many of such techniques destroy the underlying semantics of data which makes them undesirable to many real-world applications.

So, primarily this book will focus on DR techniques that preserve original data semantics. In particular, we will focus on those techniques based on rough set theory [10]. Taxonomy of DR techniques is presented in Fig. 2.1. The presented techniques are classified into two categories: those that change the underlying semantics of data during DR process and those that preserve data semantics. The choice to choose one depends upon the underlying application, e.g. if an application needs to preserve original data semantics than the DR technique to be chosen, ensure that it is preserved. However, if an application requires to discuss the relationships between attributes, then the techniques that transform the data into two or three dimensions while emphasizing these relationships may be selected.

Selection-based techniques can be both RST based or non-RST based. Besides, these there are other techniques which perform semantics-preserving dimensionality in sideline, e.g. machine learning algorithm C4.5 [11]. In this chapter we will discuss sample techniques from each of the above category. Such techniques are out of scope of this book.

2.2 Transformation-Based Reduction

These are one of the common techniques in DR literature. They perform DR process but as sideline transform the descriptive dataset features. These techniques are useful where the semantics of original features are not needed by any future process. This section discusses some of such techniques; these are classified into two categories: linear and nonlinear.

2.2.1 *Linear Methods*

Various linear methods for DR are proposed in literature with passage of time and include techniques like principal component analysis [12–16] and multidimensional scaling [17].

2.2.1.1 Principal Component Analysis (PCA)

Principal component analysis (PCA) [12, 13] is a well-known tool for data analysis and transformation and is considered the canonical means of DR. PCA is a mathematical tool that converts large number of correlated variables to smaller number of uncorrelated variables called components. The intention is to reduce the dimensions in dataset but still preserving original variability in data. The first principal component accounts for maximum of variability possible, and each of the succeeding component accounts for maximum of remaining variability.

PCA represents variance-covariance structure of high-dimensional vector with few linear combinations of the original component variables. For example, for a p-dimensional random vector $\underline{X} = (X_1, X_2, \ldots, X_p)$, PCA will find k (univariate) random variables $Y_1, Y_2, \ldots, Y_k$ called k principal components and can be defined by the following formula:

$$\begin{aligned}
Y_1 &= l_1' \ \underline{X} = l_{11}X_1 + l_{12}X_2 + \cdots + l_{1p}X_p \\
Y_2 &= l_2' \ \underline{X} = l_{21}X_1 + l_{22}X_2 + \cdots + l_{2p}X_p \\
&\vdots \\
Y_k &= l_k' \ \underline{X} = l_{k1}X_1 + l_{k2}X_2 + \cdots + l_{pk}X_p
\end{aligned}$$

Here $l_1, l_2, \ldots$, etc. coefficient vectors are chosen on the basis of the following conditions:

- First principal component = linear combination $l_1'\underline{X}$ that maximizes $\mathrm{Var}(l_1'\underline{X})$ and $\| l_1 \| = 1$
- Second principal component = linear combination $l_2'\underline{X}$ that maximizes $\mathrm{Var}(l_2'\underline{X})$ and $\| l_2 \| = 1$ and $\mathrm{Cov}(l_1'\underline{X}, l_2'\underline{X}) = 0$
- jth principal component = linear combination $l_j'\underline{X}$ that maximizes $\mathrm{Var}(l_j'\underline{X})$ and $\| l_j \| = 1$ and $\mathrm{Cov}(l_k'\underline{X}, l_j'\underline{X}) = 0$ for all $k < j$

It means that each principal component is a linear combination that maximizes variance of linear combination and has zero covariance with previous component. Figure 2.2 shows the two-dimensional normal point cloud with corresponding principal components.

Fig. 2.2 Two-dimensional normal point cloud with corresponding principal components

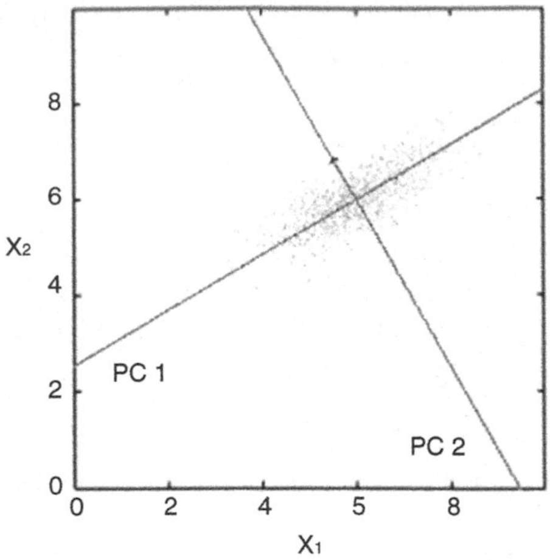

Thus PCA maximizes the variance of dataset sample vectors along their axes by locating a new coordinate system and suitably transforms the samples. The new axes are constructed in decreasing order of variance such that first variable in new axes has maximum variance and so on. Correlation in new sample space is reduced or totally removed consequently resulting in reduced redundancies. Thus DR can be performed on a dataset using PCA and then selecting appropriate number of first k principal components as per requirement and discarding the rest.

PCA, however, suffers from the following shortcomings:

- It destroys the underlying semantics of data.
- It can be used only for numeric datasets.
- It can only deal with linear projects and thus ignores any nonlinear structure in the data.
- Finally, human input is also required to decide how many of first principal components will be kept. Thus, the operator's task is to balance information loss against DR to suit the task at hand.

2.2.1.2 Projection Pursuit

Projection pursuit (PP) [18, 19] uses a quality matrix in order to project the data to lower dimensions; it opts to select interesting projects by using local optimization over projection directions of a certain "interestingness" index.

It finds the projections from high-to-low-level projection that reveal maximum of the information about the structure of the original dataset. After finding the required projections, the existing structures (clusters, surfaces, etc.) can be extracted and analysed separately.

The most basic form of PP is the scatterplot [19]. Scatterplot in its simplest form uses two dimensions to display data characteristics at a time. It is very easy to produce all two-dimensional scatterplots out of n dimensions to perform analysis. Totally there can be (2^n) pairwise scatterplots. However, this only allows the tasks to be performed only using two-dimensional scatterplots.

Hence the projections that give single dimensional projects distribution are considered to be interesting far from normal distributions. We can use different projection indices arising out of different forms of normality deviations. Friedman and Tukey in [19] opted to automate the task of projection pursuit. Numerical indices were used to describe the amount of structure presented in a projection. The heuristic search can then be applied to find the "interesting" projections using these indices. After finding a structure, it is removed from the data and data is examined for further structures and the process continues until there is no further structure to discover in the data.

Different projections may be found, each highlighting a different aspect of structure of high-dimensional data. Typically, linear projections are used due to their simplicity and interpretability.

PP has some disadvantages as well as that of PCA. PP is used for linear data, so not appropriate for nonlinear one.

2.2.1.3 Multidimensional Scaling

Multidimensional scaling (MDS) [17, 20] converts high-dimensional data to low dimensions using the distance between datapoints. So, these methods are also called distance methods; the distance, however, can be measured in terms of similarity or dissimilarity measure between datapoints. MDS, thus, converts high-dimensional data to low-dimensional data (normally these dimensions can be two to three) and tries to preserve interpoint distances up to maximum [21].

Visually, MDS maps the high dimensional data to low-dimensional data such that the patterns in the original dataset are very much present in the converted lower-dimensional space. Visually, if we draw the original space and the converted space, the points closer to each other in original space are also close to each other in converted space; similarly, the points away from each other will remain away in converted space. For example, for a given set of different brands of air fresheners, if two brands are close to each other, then MDS maps them such that they remain close to each other on converted maps.

From a slightly more technical point of view, what MDS does is to find a set of vectors in p-dimensional space such that the matrix of Euclidean distances among them corresponds as closely as possible to some function of the input matrix according to a criterion function called *stress* [20].

A simplified view of the algorithm is as follows:

1. Assign points to arbitrary coordinates in p-dimensional space.
2. Compute Euclidean distances among all pairs of points, to form the matrix.

3. Compare the matrix with the input D matrix by evaluating the stress function. The smaller the value, the greater the correspondence between the two.
4. Adjust coordinates of each point in the direction that best maximally stresses.
5. Repeat Steps 2 through 4 until stress won't get any lower.

As well as Euclidian distance is concerned, both PCA and MDS are equivalent. However, with MDS, we can use various other methods with different types of metrics and calculations. Irrespective of which method is used, basic principal remains the same. The starting point for MDS is the determination of the "spatial distance model" [20]. The following are some notations used to determine the proximities between datapoints.

Let Δ and D be two matrices of N × N dimension representing different collection of objects. The objects in the matrices are indexed by *i* and *j*, where δij is the proximity or data value of object *i* with object *j* [20] and *dij* represents the distances between pairs of points *xi* and *xj* as shown in equations below:

$$\Delta = \begin{bmatrix} \delta_{11} & \delta_{12} & \dots & \delta_{1N} \\ \delta_{21} & \delta_{22} & \dots & \delta_{2N} \\ \vdots & \vdots & \ddots & \vdots \\ \delta_{N1} & \delta_{N2} & \dots & \delta_{NN} \end{bmatrix}$$

$$D = \begin{bmatrix} d_{11} & d_{12} & \dots & d_{1N} \\ d_{21} & d_{22} & \dots & d_{2N} \\ \vdots & \vdots & \ddots & \vdots \\ d_{N1} & d_{N2} & \dots & d_{NN} \end{bmatrix}$$

The aim of MDS is to find a configuration such that the distances *dij* match, as closely as possible, the dissimilarities δij [20].

We can use different variations of MDS based on functions that transform the proximities. Classical metric multidimensional scaling is a basic form of MDS where *dij* are as close as possible to δi using Euclidian distance. This is also sometimes referred to as *principal coordinate analysis*, which is also equivalent to PCA [21]. In classical MDS methods, the relationships between δij and *dij* depend upon the metric properties of dissimilarities. Nonmetric MDS, on the other hand, refers to those methods where the relationship between δij and *dij* depends on the rank ordering of the dissimilarities [20].

In its original form, MDS [20] used the distance between datapoints and sought for the configuration that would give the similar approximation. Often a linear projection onto a subspace obtained with PCA is used. The basic idea of MDS (i.e. to approximate the distance between points into new low-dimensional subspace), however, can also be used for constructing a nonlinear projection method.

2.2.2 Nonlinear Methods

Linear DR methods are in no doubt useful, but their utility fails in case of nonlinear data. This motivated the development of nonlinear DR methods, e.g. [22–24]. An example of nonlinear method is locally linear embedding (LLE) [25, 26].

2.2.2.1 Locally Linear Embedding

LLE calculates reconstructions (embedding) which are low dimensional and neighbourhood preserving by using local symmetries of linear reconstructions (from high-dimensional data). This can be explained better by considering the following informal analogy [26]. Initial data is three dimensional, however, taking shape of rectangular manifold (two dimensional) that has been moulded to a three-dimensional S-shaped curve. Now scissors cut this manifold into small squares. Each square represents a locally linear patch of the nonlinear surface. These squares are then arranged on flat surface however by preserving angular relationships between neighbouring squares. As all transformations comprise of translation, scaling or rotation only, this is a linear mapping. Through this process algorithm uses series of linear steps to find nonlinear structure.

In the first step, it selects neighbours in datapoints. This selection can be achieved using Euclidean distance for k nearest neighbours [27]. In the second step, LLE computes the weights that linearly reconstruct datapoints using least square problem. The following cost function is used:

$$E_1(W) = \sum_{i=1}^{N} \left| \left(X_i - \sum_{j=1}^{\mathrm{K}} W_{ij} X_{\mathrm{Nj}} \right) \right|^2 \tag{2.1}$$

Finally we compute low-dimensional embedded vectors by minimizing the embedded cost function:

$$E_2(Y) = \sum_{i=1}^{N} \left| \left(Y_i - \sum_{j=1}^{\mathrm{K}} W_{ij} X_{\mathrm{Nj}} \right) \right|^2 \tag{2.2}$$

Figure 2.3 summarizes LLE algorithm.

Figure 2.4 gives the overview of these three steps.

To save the time and space, LLE also tends to accumulate very sparse matrices. It avoids dynamic programming problems as well. LLE, however, does not provide any indication about how to map a test datapoint from input space to manifold space or how to reconstruct a datapoint from its low-dimensional representation.

Similar to LLE, Laplacian Eigenmaps attempts to find low-dimensional data representation while preserving local properties of the manifold [28]. In Laplacian Eigenmaps, the local properties are based on neighbours. Laplacian Eigenmaps

1. Compute the neighbours of each data point, $\vec{X}_i$.
2. Compute the weights W_{ij} that best reconstruct each data point $\vec{X}_i$ from its neighbours, minimizing the cost in eq.(1) by constrained linear fits.
3. Compute the vectors $\vec{Y}_i$ best reconstructed by the weights W_{ij}, minimizing the quadratic form in eq. (2) by its bottom nonzero eigenvectors.

Fig. 2.3 Summary of the LLE algorithm

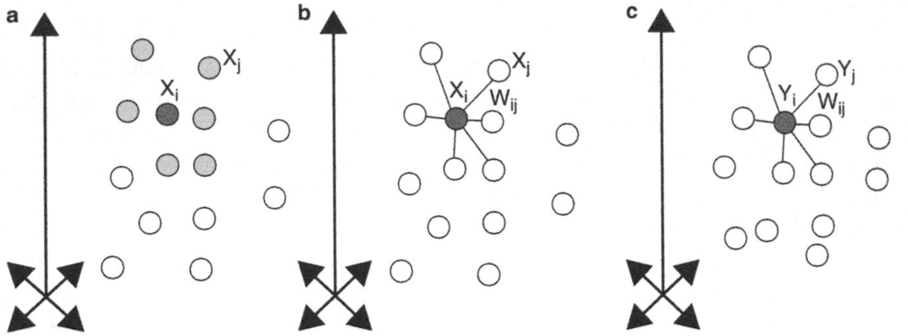

Fig. 2.4 Illustration of LLE algorithm. (**a**) Select neighbors. (**b**) Reconstruct with linear weights. (**c**) Map to embedded coordinates

minimizes the distance between a datapoint and its k nearest neighbours in an attempt to construct low low-dimensional representation of the data. Weights are used for this purpose, i.e. the distance between a datapoint and its first nearest neighbour contributes more to the cost function as compared to the distance between the datapoint and its second nearest neighbour which costs more as compared to distance between datapoint and its third neighbour and so on. Using spectral graph theory, the minimization of the cost function is defined as an eigen problem.

2.2.2.2 Isomap

Isomap [29] uses geodesic interpoint distances in contrast to Euclidean distances. The geodesic distance between the two red points is the length of the geodesic path, which is the shortest path between the points that lie on the surface [30]. Figure 2.5 shows the geodesic distance between two red points.

Isomap deals with finite datasets of points in R^n which are assumed to lie on a smooth submanifold M^d of low dimension $d < n$. From the given datapoints, algorithm tries to recover M. For a given graph G constructed out of the datapoints, Isomap attempts to find geodesic distance between datapoints in M. The Isomap algorithm consists of three basic steps:

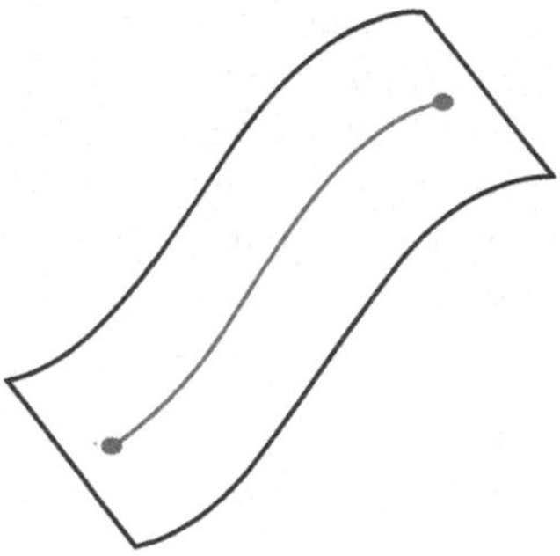

Fig. 2.5 Geodesic distance between two *red* points

1. Determine which points are neighbours on the manifold M, based on the distances between pairs of points in the input space.
2. Estimate the geodesic distances between all pairs of points on the manifold M by computing their shortest path distances in the graph *G*.
3. Apply MDS to matrix of graph distances, constructing an embedding of the data in a *d*-dimensional Euclidean space Y that best preserves the manifold's estimated geometry.

Selection of "neighbourhood" size in Isomap is critical because a large size may introduce "short circuit" into neighbourhood graph, whereas small size may make graph too sparse to approximate geodesic distance.

The success of Isomap depends on being able to choose a neighbourhood size (either ε or K) that is neither so large that it introduces "short-circuit" edges into the neighbourhood graph nor so small that the graph becomes too sparse to approximate geodesic paths accurately. Short-circuit edges occur when there are links between datapoints that are not near to each other geodesically and can lead to low-dimensional embeddings that do not preserve a manifold's true topology. The choice of neighbourhood size is an obvious limitation [29].

2.2.2.3 Multivariate Adaptive Regression Splines (MARS)

MARS [31] intends for solving regression-type problems for predicting the value of decision class from set of conditional features. Without making any assumption about underlying functional relationships, it constructs this relation using basis functions.

The basis function may be of the form

$$(x-t)_+ = \begin{cases} x-t & x>t \\ 0 & \text{otherwise} \end{cases}$$

t is the control point of a basis function and is determined from data.

The general MARS algorithm is as follows:

1. Begin with the simplest model involving only the constant basis function.
2. For each variable and possible control points, search basis functions space, and add them according to certain criteria (e.g. minimization of the prediction error).
3. Repeat Step 2 until a model of predetermined maximum complexity is obtained.
4. Apply the pruning procedure by removing the basis functions that contribute least to the overall (least squares) goodness of fit.

Brute-force approach is applied to find variables, interactions and control points, whereas least square procedure is used to determine regression coefficients. However, MARS may also generate complex mode due to noise in the data.

2.3 Selection-Based Reduction

In contrast to transformation-based techniques, which destroy the underlying semantics of data, semantics-preserving DR techniques (called feature selection) preserve original data semantics. The main intension of feature selection is to find minimal feature subset from entire dataset while maintaining a high accuracy in representing original features. Data in real work may be noisy, containing irrelevant and misleading features, so many real-world applications require FS to fix the problem. For instance, by removing these factors, learning from data techniques can benefit greatly.

2.3.1 Feature Selection in Supervised Learning

In supervised learning, feature subset selection explores feature space, generates candidate subsets and evaluates/rates them on the basis of criterion which serves as guide to search process. The usefulness of a feature or feature subset is determined by both its *relevancy* and *redundancy* [2]. A feature is relevant if it determines the value of decision feature(s); otherwise it will be irrelevant. A redundant feature is the one highly correlated with other features. Thus a good feature subset is the one highly correlated with decision feature(s) but uncorrelated with each other.

The evaluation schemes used in both supervised and unsupervised feature selection techniques can generally be divided into three broad categories [32, 33]:

1. Filter approaches
2. Wrapper methods
3. Embedded approaches.

```
FOCUS(O, c).
O, the set of all objects;
c, the number of conditional features;

(1) R ← {}
(2) for num = 1...c
(3) for each subset L of size num
(4) cons = determineConsistency(L, O)
(5) if cons == true
(6) R ← L
(7) return R
(8) else continue
```

Fig. 2.6 Pseudocode of FOCUS algorithm

2.3.2 Filter Techniques

Filter techniques perform feature selection independent of learning algorithms. Features are selected on the basis of some rank or score. A score indicating the "importance" of the term is assigned to each individual feature based on an independent evaluation criterion, such as distance measure, entropy measure, dependency measure and consistency measure [34]. Various feature filter-based feature selection techniques have been proposed in literature, e.g. [35–37]. In this section we will discuss some representative filter techniques along with advantages and disadvantages of each.

2.3.2.1 FOCUS

FOCUS [38] uses breadth-first search to find feature subsets that give consistent labelling of training data. It evaluates all the subsets of current size (initially one) and removes ones with least one inconsistency. The process continues until it finds a consistent subset or has evaluated all the possible subsets. Algorithm, however, suffers from two major drawbacks: it is very sensitive to noise or inconsistencies in training datasets and algorithm furthermore; due to exponential growth of the features power set size, algorithm is not suitable for application in domains having large number of dimensions. Figure 2.6 shows the pseudocode of FOCUS algorithm.

2.3.2.2 RELIEF

RELIEF [39] works by assigning the relevance weights to each attribute based on its ability to discern objects between different decision classes. It randomly selects

```
RELIEF(O, c, itS, ε).
O, the set of all objects;
c, the number of conditional features;
itS, the number of iterations;
ε, weight threshold value.
    (1) R ← {}
    (2) ∀ W_a, W_a ← 0
    (3) for i = 1 ... itS
    (4) choose an object x in O randomly
    (5) calculate x's nearHit and nearMiss
    (6) for j = 1...c
    (7) W_j ← W_j − diff(X_j, nearHit_j)/itS + diff(X_j, nearMiss_j)/itS
    (8) for j = 1...c
    (9) if W_j ≥ ε; R ← R ∪ {j}
    (10)        return R
```

Fig. 2.7 Pseudocode of RELIEF algorithm

an object and finds its nearHit (objects with the same class labels) and near Miss (i.e. the objects with different class labels). The distance between two objects is the sum of the number of features that differ in value between them:

$$\text{dist}(o, x) = \sum_{i=1}^{|C|} \text{diff}(o_i, x_i) \tag{2.3}$$

where

$$\text{diff}(o_i, x_i) = \begin{cases} 1, & o_i \neq x_i \\ 0, & o_i = x_i \end{cases} \tag{2.4}$$

i.e. the distance between objects is "1" if the value of an attribute differs between them and "0" if the value of an attribute is the same for both objects. Algorithm requires manual threshold value which should specify that which attribute(s) will finally be selected. The algorithm however fails to remove redundant features as two predictive but highly correlated features are both likely to be given high relevance weightings. Figure 2.7 shows the pseudocode of RELIEF algorithm.

2.3.2.3 Selection Construction Ranking Using Attribute Pattern (SCRAP)

Figure 2.8 shows the pseudocode of SCRAP [40] algorithm.

It (SCRAP [40]) performs sequential search to determine feature relevance in instance space. It attempts to identify those features that change decision boundaries in dataset by considering one object (instance) at a time; these features are considered to be most informative. Algorithm starts by selecting a random object, which is considered as first point of class change (PoC). It then selects next PoC

```
SCRAP(O).
O, the set of all objects;

(1) A ← { }; ∀ W_i, W_i = 0;
(2) T ← randomObject( ); PoC ← T
(3) while O ≠ {}
(4) O ← O − PoC; PoC_new ← NewPoC(PoC)
(5) n = dist(PoC, PoC_new)
(6) if n == 1
(7) i = diffFeature(PoC, X); A ← A ∪ {i}
(8) N ← getClosestNeighbours(PoC, n)
(9) ∀ X ∈ N
(10)        if classLabel(X) == classLabel(N)
(11)        O ← O − X
(12)        if dist(PoC, X) == 1
(13)        i = diffFeature(PoC, X); W_i = W_i − 1
(14)        else if dist(PoC, X) > 1
(15)        incrementDifferingFeatures(X, W)
(16)        R ← A
(17)        ∀ W_i, if W_i > 0 then R ← R ∪ {i}
```

Fig. 2.8 Pseudocode of SCRAP algorithm

which usually is the nearest object having different class label. After this nearest object having a different class label becomes the next PoC. These two PoCs define a neighbourhood, and dimensionality of decision boundary between the two classes is defined by the features that change between them. If only one feature changes between them, then it is considered to be absolutely relevant and is included in feature subset; otherwise their associated relevance weights (which initially are zero) are incremented. However, if objects in the same class are closer than this new PoC and differ only by one feature, then relevance weight is decremented. Objects belonging to neighbourhood are then removed, and this process continues until there is no unassigned object to any neighbourhood. Final feature subset is then selected comprising of features with positive relevance weight and those that are absolutely relevant.

Major deficiency of the approach is that it regularly chooses large number of features. This normally happens in case when weights are decremented. Feature weights remain unaffected if more than one features change between a PoC and an object belonging to same class.

2.3.3 *Wrapper Techniques*

One of the criticisms suffered by filter approaches is that the filter to select attributes is independent of the learning algorithm. To overcome this issue, wrapper approaches use classifier performance to guide the search, i.e. the classifier is wrapped in the feature selection process [44]. So, in these approaches, feature subset is selected on the basis of generalization accuracy it offers as estimated using cross validation on the training data.

Four popular strategies are [32, 41]:

1. Forward selection (FS): starting with an empty feature subset, it evaluates all features one by one, selects the best feature and combines this feature with others one by one.
2. Backward elimination (BE): initially it selects all features, evaluates by removing each feature one by one and continues to eliminate features until it selects the best feature subset.
3. Genetic search applies genetic algorithm (GA) to search feature space. Each state is defined by chromosome that actually represents a feature subset. With this representation, implementation of GA for feature selection becomes quite simple. However, the evaluation of fitness function, i.e. its classification accuracy, can be expensive.
4. Simulated annealing (SA), in contrast to GA which maintains the population of chromosomes (each chromosome represents a feature subset), considers only one solution. It implements a stochastic search as there is a chance that some deterioration in solution is accepted – this allows a more effective exploration of the search space.

Forward elimination and backward elimination terminate when adding or deleting further features do not affect classification accuracy. However these greedy search strategies do not ensure best feature subset. GA and SA can be more sophisticated approaches and can be used to explore search space in a better way.

2.3.4 Feature Selection in Unsupervised Learning

Feature selection in unsupervised learning can however be challenging because the success criterion is not clearly defined. Various unsupervised feature selection techniques have been proposed in literature, e.g. [42–44]; however, in the next section, we will discuss only a few representative techniques. Feature selection in unsupervised learning has been classified in the same way as in supervised learning. Two categories are unsupervised filters and unsupervised wrappers as discussed below:

2.3.4.1 Unsupervised Filters

The main characteristics of filter-based approaches are that features are selected on the basis of some rank or score which remains independent of the classification or clustering process. Laplacian score (LS) is one of the examples of this strategy, which can be used for DR when motivation is that the locality is preserved. The LS uses this idea for unsupervised feature selection [45]. LS selects features by preserving the distance between objects both in input and reduced output space. This criterion presumes all the features are relevant; the only thing is that they may just be redundant.

LS is calculated using a graph G that realizes nearest neighbour relationships between input datapoints. A square matrix S is used to represent G where.

$S_{ij} = 0$ unless x_i and x_j are neighbours, in which case

$$S_{ij} = e^{-\frac{||x_i - x_j|^2}{t}} \tag{2.5}$$

Here "*t*" is a bandwidth parameter. L = D – S represents Laplacian of the graph and D = degree of diagonal matrix as given below:

$$D_{ii} = \sum_j S_{ij}, \quad D_{ij,i \neq j} = 0 \tag{2.6}$$

LS can be calculated using the following calculations:

$$\tilde{m}_i = m_i - \frac{m_i^T D1}{1^T D1} 1 \tag{2.7}$$

$$LS_i = \frac{\tilde{m}_i^T L\tilde{m}_i}{\tilde{m}_i^T D\tilde{m}_i} \tag{2.8}$$

where m_i is the vector of values for the *i*th feature and 1 is a vector or 1 s of length *n*.

All the features can be scored on this criterion, i.e. how efficiently they preserve locality. This idea can be appropriate for domains where locality preservation is an effective motivation [45], e.g. image analysis. However, it may not be a sensible motivation in case of irrelevant features, e.g. in analysis of gene expression data, text classification, etc.

2.3.4.2 Unsupervised Wrappers

Wrapper-based techniques use classification or clustering process as part of feature selection to evaluate feature subsets. One such technique is proposed in [46]. Authors have used notion of a category unit (CU) [37] to present unsupervised wrapper-like feature subset selection algorithm. CU was used as evaluation function to guide the process of creating concepts and can be defined as follows:

$$CU(C,F) = \frac{1}{k}\sum_{c_i \in C}\left[\sum_{f_i \in F}\sum_{j=1}^{r_i} p\left(f_{ij}|C_l\right)^2 - \sum_{f_i \in F}\sum_{j=1}^{r_i} p\left(f_{ij}\right)^2\right] \quad (2.9)$$

Here:

$C = \{C_1, \ldots C_l, \ldots C_k\}$ is the set of clusters.
$F = \{F_1, \ldots F_i, \ldots F_p\}$ is the set of features.

CU calculates the difference between the conditional probability of a feature *i* having value *j* in cluster *l* and its prior probability. The innermost sum is over r feature values, the middle sum is over p features, and the outer sum is over k clusters. CU is used as key concept to score the quality of clustering in a wrapper-like search.

2.3.4.3 The Embedded Approach

Embedded approach is the final category of feature selection techniques. Just like construction of decision tree, feature selection in embedded approaches is an integral part of the classification algorithm. There are various techniques that perform feature selection in this category, e.g. [47–49]. Here we will discuss ESFS [50], an embedded feature selection approach which incrementally adds most relevant features. The process comprises four steps:

Step 1: Calculate belief masses of the single features.
Step 2: Evaluate single features and select initial set of potential features.
Step 3: Combine features to generate feature subsets.
Step 4: Evaluate stopping criterion and select best feature subset.

The stop criterion of ESFS occurs when the best classification rate begins to decrease while increasing the size of the feature subsets. Inspired from wrapper SFS, the algorithm incrementally selects features. It makes use of the term "belief mass" for feature processing introduced from the evidence theory, which allows to merge feature information in an embedded way.

2.4 Correlation-Based Feature Selection

Among the feature selection techniques that we have discussed till now, majority of them consider single attributes. However, an important category of the feature selection techniques is the one that considers more than one attribute (i.e. complete attribute subset) at a time while interpreting the relation between attributes themselves along with the relation between attributes and decision class. We call this category as correlation-based feature selection (CFS). CFS is based on the following assumption [51].

Good feature subsets contain features highly correlated with the class yet uncorrelated with each other.

Equation for CFS is [51]

$$r_{zc} = \frac{k\overline{r_{zi}}}{\sqrt{k + k(k-1)\overline{r_{ii}}}} \tag{2.10}$$

Here:

r_{zc} = correlation between the summed components and the outside variable.
k = number of components.
$\overline{r_{zi}}$ rzi is the average of the correlations between the components and the outside variable.
$\overline{r_{ii}}$ = average intercorrelation between components.

The following are some of the heuristics that may derive a CFS algorithm:

- Higher correlation between the components and the outside variable results in higher correlation between composite and outside variable.
- Lower intercorrelations among the components result in higher correlation between the composite and the outside variable.
- The increasing number of components in the composite increases the correlation between the composite and the outside variable.

2.4.1 Correlation-Based Measures

Mainly there are two approaches to measure the correlation between two random variables, the one based on linear correlation and the other based on information theory [52]. Under the first approach, linear correlation coefficient is one of the well-known measures. For two variables X and Y, the linear correlation coefficient is

$$r = \frac{\sum_i (x_i - \bar{x}_i)(y_i - \bar{y}_i)}{\sqrt{\sum_i (x_i - \bar{x}_i)^2}\sqrt{\sum_i (y_i - \bar{y}_i)^2}} \tag{2.11}$$

Here:

$\overline{x}_i$= mean of X.
$\overline{y}_i$ = mean of Y.

r can take the value between −1 and 1. If X and Y are fully correlated, r will either be 1 or −1 and 0 if X and Y are independent.

Linear correlations help identify redundant features and those having zero linear correlation; however, linear correlation measure cannot help in case of nonlinear correlations.

The other approach based on information theory uses the concept of information entropy. Entropy defines the measure of uncertainty of a random variable. Mathematically, entropy X of a random variable is

$$H(X) = -\sum_{i} P(x_i)\log_2(P(x_i)) \tag{2.12}$$

Entropy of X given Y is

$$H(X|Y) = -\sum_{j} p\left(y_j\right) \sum_{i} P(x_i|y_i)\log_2\left(P\left(x_i|y_j\right)\right) \tag{2.13}$$

Here:

$P(xi)$ = prior probabilities for all values of X.
$P(xi\ |yi)$ = posterior probabilities of X given the values of Y.

The amount by which the entropy of X decreases after variable Y is called information gain. Mathematically,

$$\mathrm{IG}(X|Y) = \mathrm{H}(X) - \mathrm{H}(X|Y) \tag{2.14}$$

Now we present few of the approaches based on correlation-based feature selection.

2.4.2 *Correlation-Based Filter Approach (FCBF)*

FCBF [52] used symmetrical uncertainty (SU) to evaluate the goodness of a feature subset. Algorithm is based on the following definitions and heuristics [54]:

Definition 1 (Predominant Correlation). The correlation between a feature $F_i(F_i \in S)$ and the class C is predominant iff $SU_{i,c} \geq \partial(F_i \in S)$ and $\forall F_j \in S'$ $(j \neq i)$; there exists no F_j such that $SU_{j;i} \geq SU_{i;c}$.

Definition 2 (Predominant Feature). A feature is predominant to the class, iff its correlation to the class is predominant or can become predominant after removing its redundant peers.

Heuristic 1 (if $S_{Pi}^+ = \phi$). Treat Fi as a predominant feature, remove all features in SPi ¡, and skip identifying redundant peers for them.

Heuristic 2 (if $S_{Pi}^+ \neq \phi$). Process all features in S_{Pi}^+ before making a decision on F_i. If none of them becomes predominant, follow Heuristic 1; otherwise only

```
input: S(F1, F2, ..., FN, C)     // a training data set
δ                                // a predefined threshold
output: Sbest                    // an optimal subset

begin
for i = 1 to N do begin
calculate SUi,c for Fi;
if(SUi,c ≥ δ)
append Fi to S'list;
end;
order S'list in descending SUi,c value;
Fp = getFirstElement(S'list);
do begin
        Fq = getNextElement(S'list, Fp);
        if(Fq<>NULL)
        do begin
        F'q = Fq;
        if(SUp,q ≥ SUq,c)
        remove Fq from S'list;
        Fq = getNextElement(S'list, F'p);
        else Fq = getNextElement(S'list, Fp);
        end until (Fq == NULL);
        Fp = getNextElement(S'list, Fp);
        end until (Fp == NULL);
        Sbest = S'list;
        end;
```

Fig. 2.9 Pseudocode of the FCBF algorithm

remove F_i and decide whether or not to remove features in $S_{P_i}^{-}$ based on other features in S'.

Heuristic 3 (Starting Point). The feature with the largest $Su_{i,c}$ value is always a predominant feature and can be a starting point to remove other features.

Figure 2.9 shows the pseudocode of the FCBF algorithm.

Algorithm comprises two parts. In the first parts it arranges relevant features in S'list based on predefined threshold by calculating the SU value of each feature, which are then arranged in decreasing order of their SU value. In second part it removes the redundant features from S'list by keeping only the predominant ones. Algorithm continues until there are no more features to remove. The complexity of the first part of algorithm is N, whereas that of the second part is O(N logN). Since the calculation of SU for a pair of features is linear in terms of the number of instances M in a dataset, the overall complexity of FCBF is O(MN logN).

2.4.3 *Efficient Feature Selection Based on Correlation Measure (ECMBF)*

Authors in [53] have presented an efficient correlation (between continuous and discrete measure)-based feature selection algorithm. Algorithm uses Markov blanket to determine feature redundancy. Correlation between continuous and discrete measures (CMCD) was used for feature selection.

Algorithm takes dataset, relevance threshold α and the redundancy threshold β as input. Threshold value of α is used to determine feature relevance. If the correlation between feature and class is lesser than α, it is considered to be less relevant or irrelevant and thus should be removed. If correlation between two random features is greater than β, it means one of them is redundant and thus should be removed. In that case the feature with lower correlation value is removed.

As the threshold values of threshold help select the optimal feature subset, these values should be carefully selected. Regarding the time complexity of ECMBF, it comprises of two aspects. First, the time complexity of calculating correlation between any two features which measures to be O(m*(m − 1)). Worst-case time complexity for all the features will be O(n * m2). Second, to select the relevant features, time complexity will be O(m*$\log_2$m). Thus, the overall time complexity of ECMBF will be O(n*m2). Figure 2.10 shows the pseudocode of the ECMBF algorithm.

2.5 Mutual Information-Based Feature Selection

Mutual information-based feature selection is another commonly used feature selection mechanism. Various algorithms in literature have been proposed using this measure. While correlation measures the linear or monotonic relationship between two variables. MI is more generic form to measure the decrease in uncertainty in variable X after observing variable Y. Mathematically,

$$I(X;Y) = \sum_{x,y} p(x,y) \log \frac{p(x,y)}{p(x)p(y)} \tag{2.15}$$

Here:

p(x,y) = joint probability distribution function of X and Y.
p(x) and p(y) are the marginal probability distribution functions of X and Y.

Mutual information can also be defined in terms of entropy. Here are some mathematical definitions of MI in terms of MI.

```
input: F(X1, X2, L, Xm, C)      // a training data set
α, β                  // relevance threshold and redundant threshold
output: Subset                   // an optimal subset
   (1) begin
   (2) for i = 1 to m do begin
   (3) calculate sim(Xi, C) and sim(Xi, Xj);
   (4) order features in ascending sim(Xi, C)value, and append
       Xito Srank-list;
   (5) find the greatest decent point as the relevant threshold α
       from Srank-list
   (6) if (sim(Xi, C) ≤ α)
   (7) remove Xi from Srank-list, the new subset denote as S'rank-list
       = (X1, X2, ..., Xk), for k < m;
   (8) end;
   (9) Xp = getNextElement(S'rank-list)
   (10)        do begin
   (11)        Xq = getNextElement(S'rank-list- Xp);
   (12)        if(Xq<> NULL)
   (13)        do begin
   (14)        X'q= Xq;
   (15)        if(sim(Xp, Xq) > β)
   (16)        remove Xq from S'rank-list;
   (17)        Xq = getNextElement(S'rank-list- X'q);
   (18)        else Xq = getNextElement(S rank-list - Xq);
   (19)        end until (Xq == NULL);
   (20)        Xp = getNextElement(S'rank-list -Xp);
   (21)        end until (Xp == NULL);
   (22)        Subset = S'rank-list;
   (23)        end;
```

Fig. 2.10 Pseudocode of ECMBF algorithm

$$I(X;Y) = H(X) - H(X|Y):$$

$$I(X;Y) = H(Y) - H(Y|X):$$

$$I(X;Y) = H(X) + H(Y) - H(X,Y):$$

$$I(X;Y) = H(X,Y) - H(X|Y) - H(Y|X):$$

Here H(X) and H(Y) are marginal entropies. Here we will present few MI-based feature selection algorithms.

2.5.1 A Mutual Information-Based Feature Selection Method (MIFS-ND)

MIFS-ND [54] considers both feature-feature MI and feature-class MI. Given a dataset, MIFS-NS first computes feature-feature mutual information to select the feature that has highest MI value, removes it from original set and puts in selected feature list. From the remaining feature subset, it then calculates feature-class mutual information and average feature-feature mutual information measure for each of the selected feature. Till this point, each selected feature has average feature-feature mutual information measure, and each non-selected feature has feature-class mutual information. Now from these values, it selects feature with highest feature-class mutual information and minimum feature-feature mutual information.

Algorithm then uses two terms:

- Domination count: domination count represents the number of features that a feature dominates for feature-class mutual information.
- Dominated count: dominated count represents the total number of features that a feature dominates for feature-feature mutual information.

Algorithm then selects the features that have maximum difference of domination count and dominated count. The reason behind is to select feature that is either strongly relevant or weakly redundant. The complexity of the MIFS-NS algorithm depends on the dimensionality of the input dataset. For a dataset with dimensions "d", the computational complexity of algorithm to select a subset of relevance features is $O(d^2)$. Figure 2.11 shows the pseudocode of the proposed algorithm.

Algorithm uses two major modules, Compute_FFMI which computes feature-feature mutual information and Compute_FCMI which computes feature-class mutual information. So, by using these two modules, algorithm selects the features that are highly relevant and nonredundant.

2.5.2 Multi-objective Artificial Bee Colony (MOABC) Approach

In [55] authors propose a new method for feature selection based on joint mutual information maximization (jMIM) and normalized joint mutual information maximization (NjMIM). jMIM approach applies joint mutual information and maximum of minimum approach to choose most relevant features. The intention of the proposed approach is to overcome the problem of overestimating significance of some of the features that occur during cumulative summation approximation. jMIM uses the following forward greedy search strategy (shown in Fig. 2.12) to find a feature subset of size K.

input: d, the number of features; dataset D;
$F = \{f_1, f_2, ..., f_d\}$, the set of features
output: F', an optimal subset of features
Steps:
(1) **for** i = 1 to d,**do**
(2) compute MI(f_i, C)
(3) **end**
(4) select the features f_i with maximum MI(f_i, C)
(5) $F' = F' U \{ f_i \}$
(6) $F = F - \{ f_i \}$
(7) count = 1;
(8) **while** count <= k **do**
(9) **for** each feature f_j ε F, **do**
(10) FFMI = 0;
(11) **for** each feature f_j ε F', **do**
(12) FFMI = FFMI + compute_FFMI(f_i, f_j)
(13) **end**
(14) AFFMI = Average FFMI for feature f_j.
(15) FCMI = compute_FCMI(f_i, C)
(16) **end**
(17) select the next feature f_j that has maximum AFFMI but minimum FCMI
(18) $F' = F' U \{ f_i \}$
(19) $F = F - \{ f_j \}$
(20) i = j
(21) count = count + 1;
(22) **end**
(23) Return features set F'

Fig. 2.11 Pseudocode of MIFS-ND algorithm

1. (Initialisation) Set F ← "initial set of n features"; S ← "empty set."
2. (Computation of the MI with the output class) For ∀ f_i Ɛ F compute I(C: f_i).
3. (Choice of the first feature) Find a feature f_i that maximises I(C: f_i); set F ← F\{ f_i }; set S ← { f_i }.
4. (Greedy selection) Repeat until |S| = k; (Selection of the next feature) Choose the feature $f_i = argmax_{f_i \epsilon F-S}\left(min_{f_i \epsilon S}\left(I(f_i, f_s : C)\right)\right)$; set F ← F\{ f_i }; set S ← S U { f_i }.
5. (Output) Output the set S with selected features.

Fig. 2.12 Forward greedy search

The second approach proposed in [62], i.e. NjMIM, intended to study the effect of using normalized MI instead of MI. It uses the similar goal function as that used in jMIM except that symmetrical relevance is used instead of MI. The same forward greedy search strategy was used to search feature subset for the given dataset.

2.6 Summary

In this chapter, we have provided detailed description of various feature selection techniques proposed in literature from time to time. A complete taxonomy of feature selection was presented, and feature selection algorithms according to each category in this taxonomy were explained. Efforts were made to provide the pseudocode as well along with the detailed description. Both transformation-based and selection-based techniques were discussed. This was the end of first part in which we tried to provide a strong foundation of feature selection from basic concepts to its applications in real life. In the next part, we will start with rough set theory.

Bibliography

1. Villars RL, Olofson CW, Eastwood M (2011) Big data: what it is and why you should care. White Paper, IDC. 14
2. Asuncion A, Newman D (2007) UCI machine learning repository
3. Yan J et al (2006) Effective and efficient dimensionality reduction for large-scale and streaming data preprocessing. IEEE Trans Knowl Data Eng 18(3):320–333
4. Han Y et al (2015) Semisupervised feature selection via spline regression for video semantic recognition. Neural Netw Learn Syst IEEE Trans 26(2):252–264
5. Boutsidis C et al (2015) Randomized dimensionality reduction for $ k $-means clustering. IEEE Trans Inf Theory 61(2):1045–1062
6. Cohen MB et al (2015) Dimensionality reduction for k-means clustering and low rank approximation. In: Proceedings of the forty-seventh annual ACM on symposium on theory of computing. ACM
7. Bourgain J, Dirksen S, Nelson J (2015) Toward a unified theory of sparse dimensionality reduction in euclidean space. Geom Funct Anal 25(4):1009–1088
8. Radenović F, Jégou H, Chum O (2015) Multiple measurements and joint dimensionality reduction for large scale image search with short vectors. In: Proceedings of the 5th ACM on international conference on multimedia retrieval. ACM
9. Azar AT, Hassanien AE (2015) Dimensionality reduction of medical big data using neural-fuzzy classifier. Soft Comput 19(4):1115–1127
10. Pawlak Z (1991) Rough sets, theoretical aspects about data. Springer, Dordrecht
11. Qian Y et al (2015) Fuzzy-rough feature selection accelerator. Fuzzy Sets Syst 258:61–78
12. Tan A et al (2015) Matrix-based set approximations and reductions in covering decision information systems. Int J Approx Reason 59:68–80
13. Al Daoud E (2015) An efficient algorithm for finding a fuzzy rough set reduct using an improved harmony search. Int J Mod Educ Comput Sci 7(2):16
14. Candès EJ et al (2011) Robust principal component analysis? J ACM (JACM) 58(3):11
15. Kao Y-H, Van Roy B (2013) Learning a factor model via regularized PCA. Mach Learn 91 (3):279–303
16. Varshney KR, Willsky AS (2011) Linear dimensionality reduction for margin-based classification: high-dimensional data and sensor networks. IEEE Trans Signal Process 59 (6):2496–2512
17. Der Maaten V, Laurens EP, Van den Herik J (2009) Dimensionality reduction: a comparative. J Mach Learn Res 10:66–71

18. Cunningham P (2008) Dimension reduction. Machine learning techniques for multimedia. Springer, Berlin/Heidelberg, pp 91–112
19. Friedman JH, Stuetzle W (1981) Projection pursuit regression. J Am Stat Assoc 76 (376):817–823
20. Borg I, Groenen PJF (2005) Modern multidimensional scaling: theory and applications. Springer, New York
21. Dalgaard P (2008) Introductory statistics with R. Springer, New York
22. Gisbrecht A, Schulz A, Hammer B (2015) Parametric nonlinear dimensionality reduction using kernel t-SNE. Neurocomputing 147:71–82
23. Gottlieb L-A, Krauthgamer R (2015) A nonlinear approach to dimension reduction. Discrete Comput Geom 54(2):291–315
24. Gisbrecht A, Hammer B (2015) Data visualization by nonlinear dimensionality reduction. Wiley Interdisc Rev Data Mining Knowl Disc 5(2):51–73
25. Zeng X, Luo S (2008) Generalized locally linear embedding based on local reconstruction similarity. In: Fuzzy Systems and Knowledge Discovery, 2008. FSKD'08. Fifth international conference on, vol 5. IEEE
26. Saul LK et al (2006) Spectral methods for dimensionality reduction. Semisuperv Learn: 293–308
27. Liu R et al (2008) Semi-supervised learning by locally linear embedding in kernel space. Pattern recognition, 2008. ICPR 2008. 19th international conference on. IEEE
28. Gerber S, Tasdizen T, Whitaker R (2007) Robust non-linear dimensionality reduction using successive 1-dimensional Laplacian eigenmaps. In: Proceedings of the 24th international conference on machine learning. ACM
29. Teng L et al (2005) Dimension reduction of microarray data based on local tangent space alignment. In: Cognitive informatics, 2005. (ICCI 2005). Fourth IEEE conference on. IEEE
30. Dimensionality reduction methods for molecular motion, http://archive.cnx.org/contents/02ff5dd2-fe30-4bf5-8e2a-83b5c3dc0333@10/dimensionality-reduction-methods-for-molecular-motion. Assessed on 30 Mar 2017
31. Faraway JJ (2005) Extending the linear model with r (texts in statistical science)
32. Jensen R, Shen Q (2008) Computational intelligence and feature selection: rough and fuzzy approaches, vol 8. Wiley, Hoboken
33. Cunningham P (2008) "dimension reduction". Machine learning techniques for multimedia. Springer, Berlin/Heidelberg, pp 91–112
34. Tang B, Kay S, He H (2016) Toward optimal feature selection in naive Bayes for text categorization. IEEE Trans Knowl Data Eng 28(9):2508–2521
35. Jiang F, Sui Y, Lin Z (2015) A relative decision entropy-based feature selection approach. Pattern Recogn 48(7):2151–2163
36. Singh D, Gnana AA et al (2016) Feature selection using rough set for improving the performance of the supervised learner. Int J Adv Sci Technol 87:1–8
37. Xu J et al (2013) L 1 graph based on sparse coding for feature selection. In: International symposium on neural networks. Springer, Berlin/Heidelberg
38. Almuallim H, Dietterich TG (1991) Learning with many irrelevant features. AAAI 91
39. Kira K, Rendell LA (1992) The feature selection problem: traditional methods and a new algorithm. AAAI 2
40. Raman B, Ioerger TR (2002) Instance-based filter for feature selection. J Mach Learn Res 1 (3):1–23
41. Liu H, Motoda H (eds) (2007) Computational methods of feature selection. CRC Press, Boca Raton
42. Du L, Yi-Dong Shen (2015) Unsupervised feature selection with adaptive structure learning. In: Proceedings of the 21th ACM SIGKDD international conference on knowledge discovery and data mining. ACM
43. Li J et al (2015) Unsupervised streaming feature selection in social media. In: Proceedings of the 24th ACM international on conference on information and knowledge management. ACM.

44. Singh DAAG, Balamurugan SAA, Leavline EJ (2015) An unsupervised feature selection algorithm with feature ranking for maximizing performance of the classifiers. Int J Autom Comput 12(5):511–517
45. He X, Deng C, Niyogi P (2005) Laplacian score for feature selection. NIPS 186
46. Devaney M, Ram A (1997) Efficient feature selection in conceptual clustering. ICML 97
47. Yang J, Xu H, Jia P (2013) Effective search for genetic-based machine learning systems via estimation of distribution algorithms and embedded feature reduction techniques. Neurocomputing 113:105–121
48. Imani MB, Keyvanpour MR, Azmi R (2013) A novel embedded feature selection method: a comparative study in the application of text categorization. Appl Artif Intell 27(5):408–427
49. Viola M et al (2015) A generalized eigenvalues classifier with embedded feature selection. Optimiz Lett: 1–13
50. Xiao Z et al (2008) ESFS: a new embedded feature selection method based on SFS. Rapports de recherché
51. Hall MA (2000) Correlation-based feature selection of discrete and numeric class machine learning
52. Yu L, Liu H (2003) Feature selection for high-dimensional data: a fast correlation-based filter solution. ICML 3
53. Jiang S-y, Wang L-x (2016) Efficient feature selection based on correlation measure between continuous and discrete features. Inf Process Lett 116(2):203–215
54. Hoque N, Bhattacharyya DK, Kalita JK (2014) MIFS-ND: a mutual information-based feature selection method. Expert Syst Appl 41(14):6371–6385
55. Hancer E et al (2015) A multi-objective artificial bee colony approach to feature selection using fuzzy mutual information. Evolutionary Computation (CEC), 2015 I.E. congress on. IEEE

Chapter 3
Rough Set Theory

This chapter discusses the basic preliminaries of rough set theory (RST). Since its inception, RST has been a prominent tool for data analysis due to its analysis friendly nature. RST provides a range of data structures, e.g. information systems, decision systems and approximations, to represent the real-world data. Furthermore, it provides various methods to help analyse this data. This chapter discusses the basic concepts of RST with example to set a strong foundation of RST to be used as feature selection.

3.1 Classical Set Theory

Classical set theory is a branch of mathematics that deals with collection of objects called sets. These objects are called members of that set. Rough set theory may be called an extension of classical set theory. The main problem with classical set theory is that it is concrete in nature and hence fails to model the vagueness of the universe; this is where rough set theory plays its role. Before discussing the rough set theory, here we will discuss some basics of classical set theory.

3.1.1 Sets

A set is a well-defined collection of distinct objects. "Well defined" here means that we should clearly be able to distinguish whether an object belongs to a set or not. For example, a set of even numbers between 10 and 100 is a set because every number can precisely be concluded as belonging to this set or not. On the other hand "a set of intelligent people" is not a set as there is no specific rule to specify whether a person is intelligent or not. "Distinct" means that an object will appear only once and cannot be repeated in a set.

M.S. Raza, U. Qamar, *Understanding and Using Rough Set Based Feature Selection: Concepts, Techniques and Applications*,
DOI 10.1007/978-981-10-4965-1_3

Some examples of sets are:

1. Set of support goods
2. Set of animals that lay eggs
3. Set of schools in New York
4. Set of students who scored greater than 85% score in Section II

If you note, any object can be clearly defined as belonging to these sets or not. Similarly the members in these sets will be distinct and will not be repeated.

3.1.2 *Subsets*

A set "A" will be a subset of another set "B" if every member of "A" is present in "B". Mathematically, $A \subseteq B$, here the symbol "$\subseteq$" represents the subset. We will provide details of all of the symbols used in the upcoming section.

Example
Consider the following three sets:

$$\mathrm{A} = \{2, 4, 6, 8, 10\}, \mathrm{B} = \{1, 2, 4\} \text{ and } \mathrm{C} = \{6, 8, 10\}$$

Here:

1. $C \subseteq A$, i.e. "C" is subset of "A" as all members of "C" are present in "A".
2. $B \nsubseteq A$, i.e. "B" is not subset of "A" as all members of "B" are not present in "A".
3. $A \subseteq A, B \subseteq B$ *and* $C \subseteq C$, i.e. a set is its own subset as all of its members are present in it.

3.1.3 *Power Sets*

For a set A, there exists a set whose elements are all subsets of "A". Here the elements of power subsets will not be individual objects but subsets.

Example
Consider the set A={1,2,3}:

$$\mathrm{P(A)} = \{\{\}\{1\}\{2\}\{3\}\{1,2\}\{1,3\}\{2,3\}\{1,2,3\}\}$$

P(A) comprises of all of subsets that can be formed by different arrangements of all elements of "A".

3.1.4 *Operators*

Different operators exist for manipulation of sets. Here we will discuss some basic operators.

3.1.4.1 Intersection

Sometimes we need the elements common in all sets. Intersection operator is intended for this purpose. By definition, intersection of two sets "A" and "B" is another set "C" that contains all the elements common in "A" and "B". Intersection is denoted by the symbol "∩". Mathematically, $A \cap C = \{x \mid (x \in A)\ and\ (x \in B)\}$.

Example
A = {1,2,3,4,5,6}
B = {4,5,6,7,8,9}
C = {0,1,2,3}

1. $A \cap B = \{4,5,6\}$
2. $A \cap C = \{1,2,3\}$
3. $B \cap C = \{\varnothing\}$

Note: The symbol "∅" represents an empty set.

3.1.4.2 Union

Union of two sets "A" and "B" is another set "C" that contains all the elements of "A" and "B" such that if an element appears in both sets, it is taken once. Union is denoted by symbol "∪". Mathematically, $A \cup C = \{\ x \mid (x \in A)$ or $(x \in B)\}$, i.e. to belong to a union of sets "A" and "C", an element should be a member of either of "A" or "B".
A = {1,2,3,4,5,6}
B = {4,5,6,7,8,9}
C = {0,1,2,3}

1. $A \cup B = \{1,2,3,4,5,6,7,8,9\}$
2. $A \cup C = \{0,1,2,3,4,5,6\}$
3. $B \cup C = \{0,1,2,3,4,5,6,7,8,9\}$

Note that "4,5,6" appear both in "A" and "B"; however, in $A \cup B$ they are written once.

3.1.4.3 Complement

The complement of “A” in “B” is a set “C” that contains the elements present in “A” but not in “B”. It is denoted by “A – B” or “A \ B”. Normally we consider all the sets to be subsets of universal set “ $\mathbb{U}$ ”. So, $\mathbb{U}$ A becomes the absolute complement of “A”.
A = {2,3,4}
B = {0,1,2,3}

1. $A - B = \{4\}$
2. $B - A = \{0, 1\}$

3.1.4.4 Cardinality

Cardinality of a set is the total number of objects present in that set, e.g. symbolically cardinality of set “A” will be represented as “|A|”. Consider the following set:

A = {2,4,6,8,10}
A = {1,3,5}
|A| = 5
|B| = 3

3.1.5 *Mathematical Symbols for Set Theory*

Here is a list of symbols used in set theory taken from [1]. Memorizing these symbols is helpful as they are also used in rough set theory (Table 3.1).

3.2 Knowledge Representation and Vagueness

Knowledge is the ability to classify objects. An object here simply refers to its classical definition that is “an object is abstraction or realization of real-world entities that show some properties”. The properties are represented in the form of attributes. We have already discussed attributes in detail in the first chapter.

This collection of objects is called universe; they form sets called universe of discourse. Examples may be set of students, set of medicines, set of support goods, etc. Classification of objects in simple words refers to finding subsets of the universe having objects that have common values regarding the concept under consideration, e.g. from a set of fruits having different fruits, and classifying the objects that belong to the same season or have the same colour. The clearest and unambiguous classification is when an object belongs to a single classification under the single concept. For example consider Table 3.2.

Table 3.1 Mathematical symbols and their meanings [1]

Symbols	Description	Symbol	Description
{ }	Set	A×B	Cartesian product
A ∩ B	Intersection	|A|	Cardinality
A ∪ B	Union	#A	Cardinality
A ⊆ B	Subset	$\aleph_0$	Aleph-null
A ⊂ B	Proper subset/strict subset	$\aleph_1$	Aleph-one
A ⊄ B	Not subset	Ø	Empty set
A ⊇ B	Superset	$\mathbb{U}$	Universal set
A ⊃ B	Proper superset/strict superset	$\mathbb{N}_0$	Natural numbers/whole numbers set (with zero)
A ⊅ B	Not superset	$\mathbb{N}_1$	Natural numbers/whole numbers set (without zero)
2A	Power set	$\mathbb{Z}$	Integer numbers set
$\mathcal{P}(A)$	Power set	$\mathbb{Q}$	Rational numbers set
A = B	Equality	$\mathbb{R}$	Real numbers set
A^c	Complement	$\mathbb{C}$	Complex numbers set
A\B	Relative complement	a∈A	Element of
A − B	Relative complement	x∉A	Not element of
A Δ B	Symmetric difference	(a,b)	Ordered pair
A ⊖ B	Symmetric difference		

Table 3.2 Set of persons

Person	Name	Status
X_1	John	Student
X_2	Elizbeth	Faculty
X_3	David	Faculty
X_4	Yushra	Student
X_5	Peter	Student
X_6	Alia	Faculty
X_7	Nicson	Student

Now if we classify the above set of universe w.r.t to "status", we will have two sharp classifications as follows (Fig. 3.1):

But what if David is also a student taking some professional course? Will it belong to "student" classification or "faculty"? This example also shows the vagueness we have in our real world. Unfortunately classical set theory fails to represent this vagueness. It is concrete in nature; here an object can either belong to a set or not; representing information in this way, much of the information is lost. For example, consider the set:

$$A = \{\text{students with CGPA} \geq 3.5\} = \{\text{John, Peter, Elizbeth}\}$$

By means of classical set theory, both "John" and "Peter" are equally competent (w.r.t. grades); however, the fact that "Peter" has more grades than "John" cannot

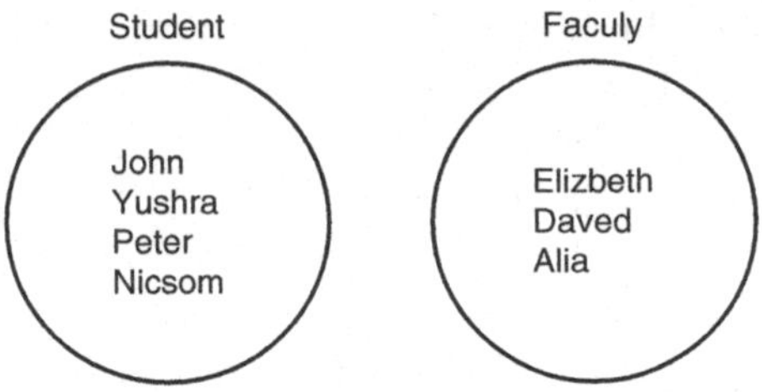

Fig. 3.1 Classification of objects given in Table 3.2

be formulated here. Some extensions of classical set theory have proposed including rough set theory, fuzzy set theory and its hybridization with rough set theory, i.e. fuzzy-rough set theory. Complete description of fuzzy set theory is out of scope of this book; however, in the next chapter, we will give some introduction of it to discuss fuzzy-rough set theory.

3.3 Rough Set Theory (RST)

As proposed by Zdzislaw Pawlak [2], RST has become a topic of great interest over the past 10 years and has been successfully applied to many domains by researchers. As discussed above, classical sets are concrete in nature, so these fail to model the vagueness of real world. RST overcomes the problem by the concept of set approximations (discussed later). RST is a combination of data structures and tools/techniques that make it analysis friendly. Here we will discuss some preliminaries of RST.

3.3.1 Information Systems

An information system is just like a flat table or view [2] comprising of objects and their attributes. An IS (Λ) is defined by a pair (U,A) [2] as given below:

$$\Lambda = (U, A)$$

Here:

U = finite nonempty set of objects
A = attributes of the objects

Every attribute $a \in A$ has a value set represented by V_a as shown below. Each value set of an attribute contains all possible values of that attribute.

Table 3.3 Information system

Customer	Age	Income
X_1	35–40	30,000–40,000
X_2	35–40	30,000–40,000
X_3	40–45	50,000–60,000
X_4	25–35	20,000–30,000
X_5	40–45	50,000–60,000
X_6	25–35	2000–30,000
X_7	25–35	20,000–30,000

$$a : U \rightarrow V_a$$

Table 3.3 presents an information system $\Lambda = (U, A)$ where

$$U = \{x_1, x_2, x_3, x_4, x_5, x_6, x_7\}$$
$$A = \{Age, Incom\}$$

3.3.2 *Decision Systems*

Decision systems (DS) [2] are a special form of information system having decision attribute also called the class of the object. Every object belongs to a specific class. The value of the class depends on other attributes called conditional attributes. Formally,

$\alpha = (U, C \cup \{D\})$

where:

C = set of conditional attributes
D = decision attribute (or class)

Table 3.4 shows a decision system with policy as decision attribute (or class).

3.3.3 *Indiscernibility*

A decision system represents all knowledge about a model. This table may be unnecessarily large by two ways: there may be identical or indiscernible objects having more than one occurrence and there may be superfluous attributes. The notion of equivalence is recalled first. A binary relation is called equivalence relation if it is reflexive, i.e. an object is in relation with itself xRx; symmetric, i.e. if xRy, then yRx; and transitive, i.e. if xRy and yRz, then xRz. The equivalence class of an element consists of all objects such as xRy.

Table 3.4 Decision system

Customer	Age	Income	Policy
X_1	35–40	30,000–40,000	Platinum
X_2	35–40	30,000–40,000	Platinum
X_3	40–45	50,000–60,000	Gold
X_4	25–35	20,000–30,000	Silver
X_5	40–45	50,000–60,000	Gold
X_6	25–35	20,000–30,000	Silver
X_7	25–35	20,000–30,000	Gold

Let $A = (U, C \cup \{D\})$ be a decision system; indiscernibility defines an equivalence relation between objects in A. For any $c \in C$ in A, there exists an indiscernibility relation $IND_A(C)$:

$$IND_A(C) = \{(O_1 = O_2) \in U^2 \mid \forall c \in C\; c(O1) = c(O2)\} \tag{3.1}$$

$IND_A(C)$ (also denoted by $[x]_c$) is called a "C-indiscernibility" relation. If two objects $(O_1, O_2) \in IND_A(C)$, then these objects are indiscernible or indistinguishable w.r.t. C. Considering Table 3.3, objects x_1,x_2 are indiscernible w.r.t. attribute "age". Similarly objects x_3 and x_5 are indiscernible w.r.t. attribute "income". The subscript is normally omitted if we are sure about which information system is meant. In Table 3.3,

$$IND(\{Age\}) = \{\{x_1, x_2\}, \{x_3, x_4\}, \{x_5, x_6, x_7\}\}$$
$$IND(\{Income\}) = \{\{x_1\}, \{x_2\}, \{x_3, x_4\}, \{x_5, x_6, x_7\}\}$$

3.3.4 Approximations

Most of the sets cannot be identified unambiguously, so we use approximation. For an information system where$B \subseteq A$, we can approximate the decision class X by using the information contained in B. The lower and upper approximations are defined as follows [2]:

$$X : \underline{B}X = \{x \mid [x]_B \subseteq X\} \tag{3.2}$$

$$X : \bar{B}X = \{x \mid [x]_B \cap X \neq 0\} \tag{3.3}$$

Lower approximation defines the objects that are definitely members of X with respect to the information in "B". Upper approximation on the other hand contains objects that with respect to "B" can possibly be members of "X". The boundary region defines the difference between lower and upper approximation.

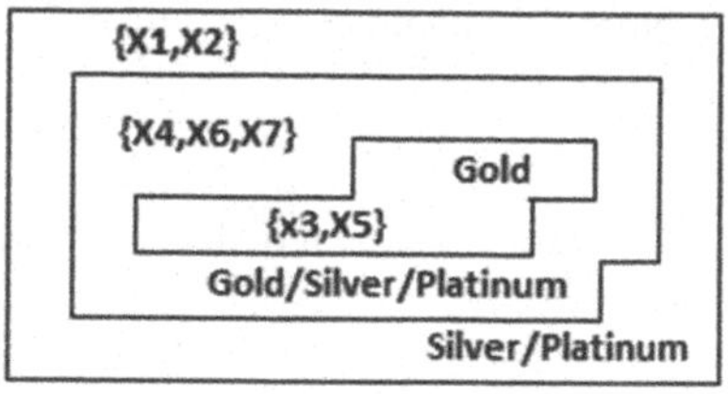

Fig. 3.2 Approximation diagram

$$X : BN_B(X) = \bar{B}X - \underline{B}X \tag{3.4}$$

Using the decision system shown in Table 3.1(b), the situation can be sketched as in Fig. 3.2. The lower boundary of "policy" defines all the equivalence classes that can surely belong to class policy=gold. Upper boundary defines classes that can possibly belong to policy=gold.

$$\underline{B}Policy = \{\{X_3, X_5\}\}$$
$$\bar{B}Policy = \{\{X_3, X_5\}, \{X_4, X_6, X_7\}\}$$

The boundary region is $\bar{B}Insurance - \underline{B}Insurance = \{\{X_4, X_6, X_7\}\}$. As it is nonempty, so the set is a rough set.

3.3.5 Positive Region

Lower approximation is also called positive region. Let P and Q be equivalence relations over U; then the positive region can be defined as

$$POS_P(Q) = \bigcup_{X \in \mathbb{U}/Q} \underline{PX} \tag{3.5}$$

where P is the set of conditional attributes and Q is the decision class. The positive region is the union of all equivalence classes in $[X]_P$ that are subset of (or are contained by) target set.

Considering Table 3.1(b), we calculate positive region for set "policy=gold" as follows:

First we will calculate $[X]_P$. Here,

$$P_1 = \{x_1, x_2\}$$
$$P_2 = \{x_3, x_5\}$$
$$P_3 = \{x_4, x_6, x_7\}$$

Now we calculate $[X]_Q$ where Q implies the concept "insurance=gold". Here,

$$Q = \{x_3, x_5, x_7\}$$

It means we cannot distinguish between x_3, x_5 and x_7 with respect to information contained in Q.

Here for concept "policy = gold", only P_2 class belongs to Q. So, positive region for Q will be

$$POS_P(Q) = \{x_3,\ x_5\}.$$

3.3.6 *Discernibility Matrix*

A discernibility matrix M of elements m_{ij} such that each element defines the set of attributes where the object $x_i, x_j \in U$ differs to each other. For an information system $A = (U, C \cup \{D\})$, the discernibility matrix is a n x n matrix with elements m_{ij} as given below.

$$x_{ij} = \{\ a \in A\ |a(xi) \neq a(xj)\}\ where\ i,j = 1,2,3,\ \ldots, n \quad (3.6)$$

Discernibility matrix is also a tool to find "reducts". A reduct is a subset of features that provide the same partition of universe as obtained by the entire set of features. We will have in-depth discussion on reducts in upcoming topics. We will now explain discernibility matrix with an example; consider the following decision system given in Table 3.5.

The discernibility matrix for this decision system is given in Table 3.6.

Consider the entry m31 (just leave the leftmost column and topmost row as these are for heading purpose only), i.e. the entry at intersection of objects x3 and x1; the value is (T∨I∨L), which means that objects x1 and x3 are discernible w.r.t. attributes "T","I" or "L". Note that diagonal is empty; otherwise it will repeat the entries.

Table 3.5 A sample decision system

	T	I	P	L	d
x1	1	1	1	2	1
x2	1	0	1	0	0
x3	2	0	1	1	0
x4	1	2	1	0	1
x5	1	1	1	0	0
x6	1	2	1	2	1
x7	1	2	0	1	1
x8	2	0	0	2	0

Table 3.6 Discernibility matrix

	X1	X2	X3	X4	X5	X6	X7	X8
x1	φ							
x2	(I∨L)	Φ						
x3	(T∨I∨L)	(T∨L)	Φ					
x4	(T∨I∨L)	(T∨I)	(T∨I∨L)	φ				
x5	(T∨L)	(T∨I)	(T∨I∨L)	(I)	φ			
x6	(T∨I)	(T∨I∨L)	(T∨I∨L)	(L)	(I∨L)	φ		
x7	(I∨P∨L)	(I∨P∨L)	(T∨I∨P)	(T∨P∨L)	(T∨I∨P∨L)	(T∨P∨L)	φ	
x8	(T∨I∨P)	(T∨P∨L)	(P∨L)	(T∨I∨P∨L)	(T∨I∨P∨L)	(T∨I∨P)	(T∨I∨L)	φ

3.3.7 *Discernibility Function*

A discernibility function f_A for an information system A is a Boolean function such that

$$f_A = \cap\{\cup c_{ij} | c_{ij} \neq \varnothing\} \tag{3.7}$$

Discernibility matrix combined with discernibility function helps us in defining reducts and rule extraction. We will see the example of rule extraction in the upcoming section. Consider the information system given in table above, the discernibility function for this system can be defined as follows:

$f_A(T, I, P, L)$ = (I∨L)(T∨I∨L) (T∨I∨L) (T∨L)(T∨I) (I∨P∨L) (T∨I∨P)
(T∨L) (T∨I) (T∨I) (T∨I∨L) (I∨P∨L) (T∨P∨L)
(T∨I∨L) (T∨I∨L) (T∨I∨L) (T∨I∨P) (P∨L)
(I) (L) (T∨P∨L) (T∨I∨P∨L)
(I∨L) (T∨I∨P∨L) (T∨I∨P∨L)
(T∨P∨L) (T∨I∨P)
(T∨I∨L)

After solving, the function simplifies to IL (i.e. $I \cap L$). There is a relation between discernibility function and discernibility matrix; each row in discernibility function corresponds to a column in discernibility matrix. The entries in discernibility function discern the objects from others, e.g. the second last row indicates that object x7 differs from x8 w.r.t. attributes T, I and L. A; discernibility function based on column K from discernibility matrix is called K-discernibility function and thus identifies the minimum set of reducts required to discern X_k from other objects.

3.3.8 *Decision-Relative Discernibility Matrix*

Decision-relative discernibility matrix is a special form of discernibility matrix $M^d(A) = c_{ij}^d$ assuming $c_{ij}^d = \varnothing if\ d(x_i) = d(x_j)$ and $c_{ij}^d = c_{ij}\ otherwise$. As we construct

Table 3.7 Decision system given in Table 3.6 rearranged according to decision class

	T	I	P	L	d
x1	1	1	1	2	1
x4	1	2	1	0	1
x6	1	2	1	2	1
x7	1	2	0	1	1
x2	1	0	1	0	0
x3	2	0	1	1	0
x5	1	1	1	0	0
x8	2	0	0	2	0

Table 3.8 Decision-relative discernibility matrix

	X1	X4	X6	X7	X2	X3	X5	X8
x1	Φ							
X4	Φ	Φ						
X6	φ	Φ	Φ					
X7	φ	φ	Φ	Φ				
X2	I,L	T,I	T,I,L	I,P,L	Φ			
X3	T,I,L	T,I,L	T,I,L	T,I,P	Φ	Φ		
X5	T,L	I,	I,L	T,I,P,L	Φ	Φ	Φ	
x8	T,I,P	T,I,P,L	T,I,P	T,I,L	φ	Φ	φ	Φ

the discernibility function from discernibility matrix, similarly decision-relative discernibility function can be constructed from decision-relative discernibility matrix. Simplification of this function results in set of all the decision relative reducts of A.

We will now explain this with an example. We will consider the same information system given in the table above. However, we have rearranged the rows to order them according to decision class. The resulting discernibility matrix is shown in the table below. It is a symmetrical matrix with empty diagonal. Similarly all the entries where decision is equal are also empty (Tables 3.7 and 3.8).

Table 3.8 Decision-relative discernibility matrix for decision system shown in Table 3.7

From the definition of decision-relative discernibility matrix, it follows that considering one column gives us discernibility function that can be used to discern that object from others. For example, consider the column of X1; it gives the discernibility function that can be used to discern object X1 from others. Figures 3.3, 3.4, 3.5, and 3.6 (taken from [3]) show four types of indiscernibility.

Figure 3.3 shows indiscernibility relation not related to a particular concept and decision attribute. These reducts are minimal subsets that discern all the cases from each other up to some extent.

Figure 3.4 shows indiscernibility relation relative to a decision attribute but not relative to a particular case. These reducts are minimal set of attributes that provide the same classification as obtained by the entire set of conditional attributes.

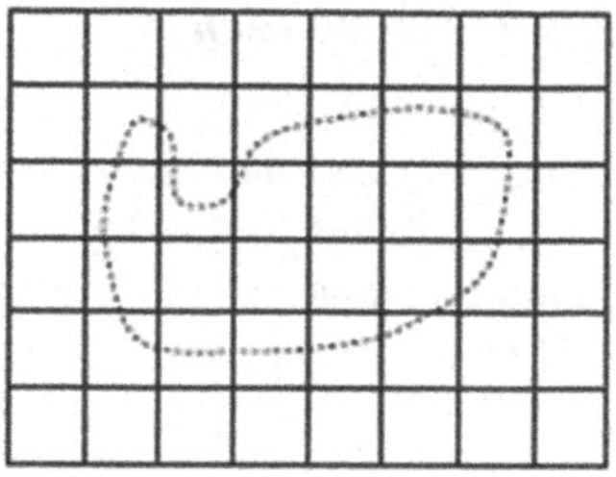

Fig. 3.3 Indiscernibility relation not related to a particular concept and decision attribute

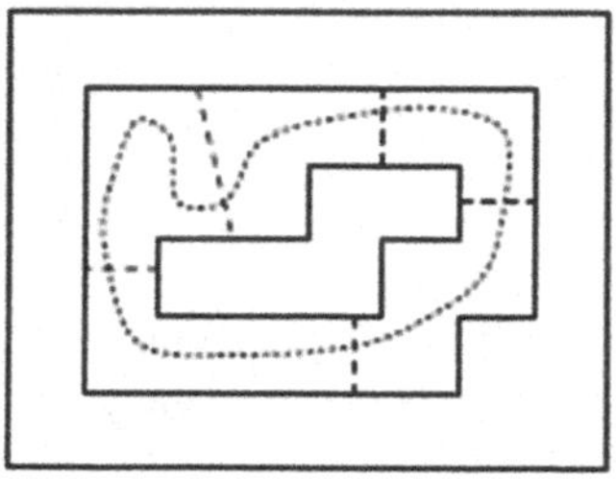

Fig. 3.4 Indiscernibility relation related to a decision attribute but not relative to a particular case

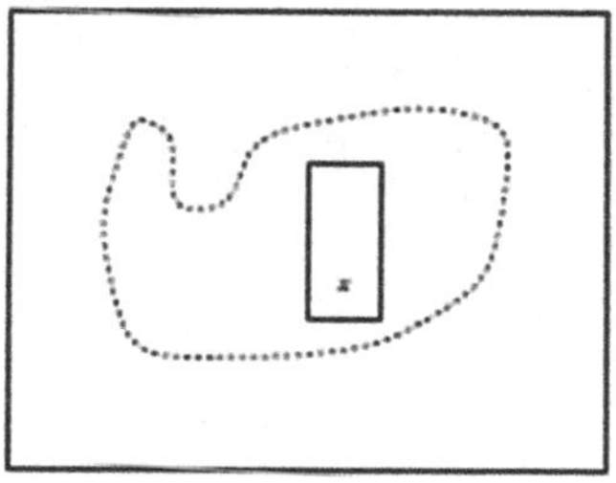

Fig. 3.5 Indiscernibility relation related to a case or object X but not relative to decision attribute

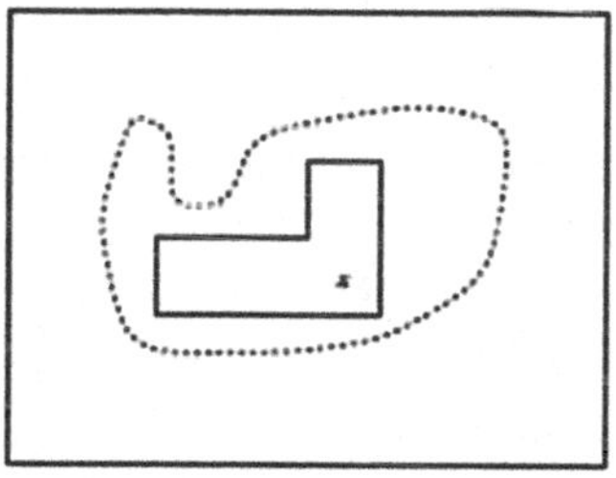

Fig. 3.6 Indiscernibility relation related both to a case and decision attribute

Figure 3.5 shows indiscernibility relation relative to a case or object X but not relative to decision attribute. Reducts of such type are minimum subset of attributes that can discern object X from others up to the same extent as done by a full set of conditional attributes.

Figure 3.6 shows indiscernibility relation relative to both a case and decision attribute. These reducts let us determine the outcome of a case as determined by full set of conditional attributes.

3.3.9 Dependency

Dependency defines how uniquely the value of an attribute determines the value of other attributes. Dependency defines how uniquely the value of an attribute determines the value of other attributes. An attribute "D" depends on the other attribute "C" by degree "K" calculated by

$$k = \gamma(C, D) = \frac{|POS_C(D)|}{|U|} \tag{3.8}$$

where

$$POS_C(D) = \cup_{X \in U/D} \underline{C}(X) \tag{3.9}$$

is called positive region of "U/D" w.r.t. "C" as discussed in Sect. 1.5. "K" is called the degree of dependency and specifies the ratio of the elements that can positively be contained by partition induced by D, i.e. U/D. If K = 1, D fully depends on C; for 0 <K< 1, D depends partially on C; and for K = 0, D does not depend on C. It is clear that if K=1, i.e. D totally depends on C, then IND(C) ⊆ IND (D); in simple words the U/C is finer than U/D.

Finally dependency is calculated as follows:

$$k = \gamma(C, D) = \frac{|POS_C(D)|}{|U|} \tag{3.10}$$

Calculating dependency using positive region requires three steps.

1. First construct the equivalence class structure using decision classes.
2. Construct equivalence class structure using current attribute set.
3. Calculate positive region:

Here we provide details of each of these steps. Consider the decision system $DS = \{\{State, Qualification\} \cup \{Job\}\}$ given in Table 3.9:

We will calculate $k = \gamma(\{state, Qualification\}, Job)$ using positive region-based approach.

Step 1

First step is calculating positive region-based dependency measure to calculate equivalence classes using decision attribute ("job" in our case).

Equivalence class structure specifies all the indiscernible objects, i.e. the objects w.r.t. given attributes cannot be distinguished. In our case we will have two equivalence classes as follows:

Table 3.9 Sample decision system

U	State	Qualification	Job
x_1	S1	Doctorate	Yes
x_2	S1	Diploma	No
x_3	S2	Masters	No
x_4	S2	Masters	Yes
x_5	S3	Bachelors	No
x_6	S3	Bachelors	Yes
x_7	S3	Bachelors	No

$$Q1 = \{x_1, x_4, x_6\}$$
$$Q2 = \{x_2, x_3, x_5, x_7\}$$

Note that if we consider the value of "job" as "yes", we cannot distinguish among x_1, x_4 and x_6.

Step 2
After calculating the equivalence classes using decision attribute, next step is to calculate equivalence class structure for decision attributes (in our case "{State, Qualification"}). Calculating equivalence classes using conditional attributes requires comparison of value of each attribute for each record to find indiscernible objects. The equivalence classes in our case will be:

$$P1 = \{x1\}$$
$$P2 = \{x2\}$$
$$P3 = \{x3, x4\}$$
$$P4 = \{x5, x6, x7\}$$

Step 3
Positive region specifies which equivalence classes in Step 2 are contained by or subset of equivalence classes identified in Step 1. First we will check which classes from P1 to P4 are subset of Q1, and then we will calculate which classes from P1 to P4 are subsets of Q2. This process will be used for all classes in Step 2, and we will identify all classes that are subset of equivalence classes in Step 1.

Here:

$$P_1 \subseteq Q_1$$
$$P_2 \subseteq Q_2$$

No other class from P_1, P_2, P_3 and P_4 is subset of either of Q_1 and Q_2. So the dependency will be

$$k = \gamma(C,D) = \frac{|POS_C(D)|}{|U|} = \frac{|P_1| + |P_2|}{|U|}$$

$$k = \gamma(\{State, Qualification\}, Job) = \frac{2}{7}$$

This process will take a considerable amount of time for datasets with large numbers of attributes and instances. Thus, this factor makes a positive region-based dependency measure a bad choice for use in feature selection algorithms against these datasets.

3.3.10 Reducts and Core

One way of dimensional reduction is keeping only those attributes that preserve the indiscernibility relation, i.e. classification accuracy. Using a selected set of attributes provides the same set of equivalence classes that can be obtained by using the entire attribute set. The remaining attributes are redundant and can be reduced without affecting classification accuracy. There are normally many subsets of such attributes called reducts. Mathematically reducts can be defined using the dependency as follows:

$$\gamma(\mathrm{C},\mathrm{D}) = \gamma(\mathrm{C}',\mathrm{D}) \text{ for } \mathrm{C}' \subseteq \mathrm{C} \tag{3.11}$$

i.e. an attribute set C' $\subseteq$ C will be called reduct w.r.t. D, if the dependency of D on C' will be the same as that of its dependency on C.

Calculating the reducts comprises of two steps. First, we calculate dependency of the decision attribute on entire dataset. Normally this is "1"; however, for inconsistent datasets, this may be any value between "0" and "1". In the second step we try to find the minimum set of attributes on which decision attribute has the same dependency value as that of its value on the entire set of attributes. In this step we may use any rough set-based feature selection algorithm. It should be noted that there may be more than one reduct sets in a single dataset.

We will now explain it with the help of an example. Consider Table 3.10.

For our first step, we calculate the dependency of decision attribute "D" on conditional attributes C={a,b,c}. Here we find that

$$\gamma(C,D) = 1$$

For the second step, we have to find the attribute subsets such that condition mentioned in Eq. (3.8) is satisfied. Here we see that we may have two subsets that satisfy the condition.

Table 3.10 Sample decision system

U	a	b	c	D
X_1	1	1	3	x
X_2	1	2	2	y
X_3	2	1	3	x
X_4	3	3	3	y
X_5	2	2	3	z
X_6	1	1	2	x
X_7	3	3	1	y

$$\gamma(\{a,b\},D) = 1$$
$$\gamma(\{b,c\},D) = 1$$

Representing them with R_1 and R_2,

$$R_1 = \{a,b\}$$
$$R_2 = \{b,c\}$$

So either of R1 and R2 provide the same classification accuracy as provided by entire of the conditional attribute set thus can be used to represent entire dataset. It is important that reduct set should be optimal, i.e. it should contain a minimum number of attributes to better realize its significance; however, finding optimal reduct is a difficult task as it requires exhaustive search with more number of resources. Normally exhaustive algorithms are used to find reducts in smaller datasets; however, for datasets beyond smaller size, the other category of algorithms, i.e. random or heuristics-based search, is used, but the drawback of these algorithms is that they do not produce optimal result. So getting the optimal reducts is a trade-off between the resources and reduct size.

Core is another important concept in rough set theory. Normally the reduct set is not unique in a dataset, i.e. we may have more than one reduct set. Although reduct may contain the same amount of information otherwise represented by the entire attribute subset, even in reduct there are attributes that are more important than others, i.e. these attributes cannot be removed without affecting the classification accuracy of the reducts. Mathematically it can be written as $Core = \bigcap_{i=1}^{n} R_i$ where R_i is *i*th reduct set. So, core is the attribute or set of attributes common to all reduct sets. In our example explained above, it is clear that the attribute {b} is common in all reduct sets, so {b} is the core attribute here. Manually it can be seen that removing attribute {b} from either of the reducts affects dependency of decision class on the rest of the attributes in that reduct, thus affecting the classification accuracy of the reduct.

3.4 Discretization Process

Discretization is the process of converting or partitioning continuous attributes, features or variables to discretized or nominal attributes/features/variables/intervals. Discretization process determines how coarsely we see the world, e.g. student marks can be any real number between zero (0) and hundred (100) in a subject, but we normally convert it to three to four grades. Another example may be blood pressure. Although the value is already measured in discrete (natural number), however it may be determined in three intervals. So it is apparent that discretization of features is a complex step; however, it is essential especially from feature selection point of view as the data in real life may be continuous. So before performing feature selection, we have to apply the discretization step to convert the attribute values to discrete values.

Various techniques have been proposed in literature for discretization [46, 68, 91, 232, 287, 354]. Here we will present a simple RST-based discretization method taken from [3]. In discretization of basic decision system $\Delta = (U, A \cup \{d\})$, where $V_a = [v_a, \omega_a)$ is an interval of real numbers, we try to find partition P_a of V_a for $a \in A$. Any partition of V_a is defined by a sequence of cuts $V_1 < V_2 < \ldots < V_n$ from V_a. So in discretization we identify some family of partitions by finding the cuts that satisfy natural condition.

The following is a sequence of steps in a RST discretization process taken from [3].

Step 1: For each attribute, create a set c(U) where $a \in A$.
Step 2: Determine $V_a = [v_a, \omega_a)$ for each attribute $a \in A$.
Step 3: Define intervals on value sets defined in Step 2.
Step 4: Create a set of cuts P for each interval.
Step 5: Discretize the attribute a to new attribute a^P using the cuts.

The process steps are defined at abstract level; now we explain all of these steps with the help of an example.

Example
Consider Table 3.11. It comprises two conditional attributes and one decision attribute. Conditional attributes "a" and "b" have real values.

Table 3.11 A decision system comprising real numbers

A	a	b	d
u1	0.8	2	1
u2	1	0.5	0
u3	1.3	3	0
u4	1.4	1	1
u5	1.4	2	0
u6	1.6	3	1
u7	1.3	1	1

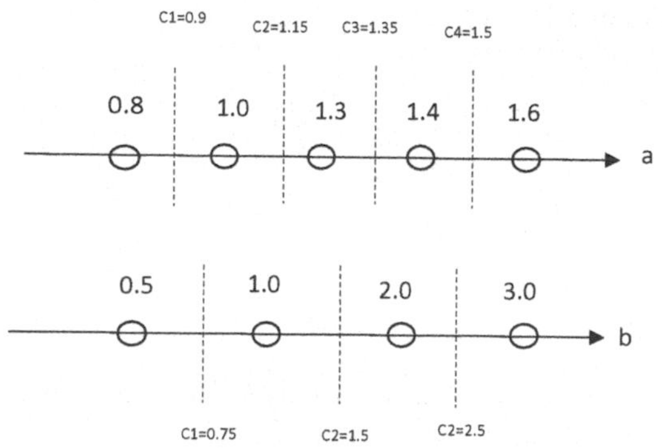

Fig. 3.7 Graphical representation of intervals and cuts

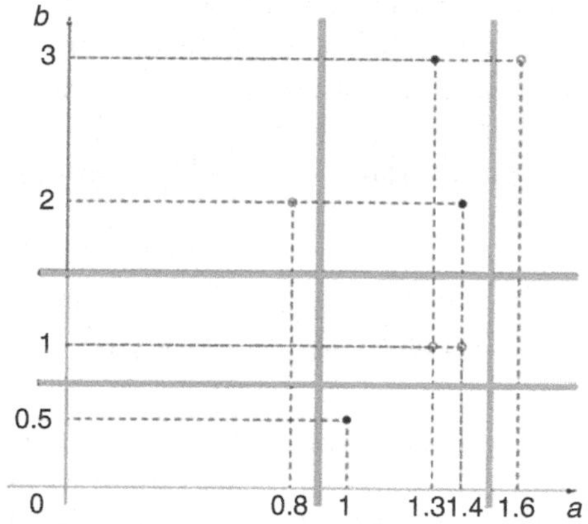

Fig. 3.8 Graphical representation of relation between data and cuts

The first step is to determine $V_a = [v_a, \omega_a)$ for each attribute $a \in A$. The values of attributes "a" and "b" for universe U are given by the sets

a(U) = {0.8,1,1.3,1.4,1.6} and b(U) = {0.5,1,2,3}

Here we have two attributes, so $V_a = [0.8,2)$ and $V_b = [0.5,4)$.

The next step is to define intervals based on the value set of conditional attributes, in our case,

For attribute "a": [0:8:1); [1; 1:3); [1:3; 1:4); [1:4; 1:6)
For attribute "b": [0:5; 1); [1; 2); [2; 3)

The next step is to form cuts. A cut is a pair (a,c) where $a \in A$. A cut may simply be a middle point of the intervals defined above. So,

In case of attribute "a": P_a = (a; 0:9); (a; 1:15); (a; 1:35); (a; 1:5)
In case of attribute "b": P_b = (b; 0:75); (b; 1:5); (b; 2:5)

Figure 3.7 shows the graphical representation of intervals and cuts.

Figure 3.8 shows the relation between data and cuts. These sets of cuts define new attribute a_p for any value of a. This mechanism works as follows:

Table 3.12 Discretized decision system

A	a	b	d
u1	0	2	1
u2	1	0	0
u3	1	2	0
u4	1	1	1
u5	1	2	0
u6	2	2	1
u7	1	1	1

The value of attribute may fall anywhere in between values of cuts. We suppose the cut set P_a = (a; 0.9); (a; 1.15); (a; 1.35); (a; 1.5); now any value from V_a that falls below 0.9 will be given value of "1", any value that will fall in range between 0.9 and 1.5 will be given value of "1", any value in range 1.15–1.35 will be given value of "2", and so on. Similarly for P_b = (b; 0.75); (b; 1.5); (b; 2.5), any value below 0.75 will be assigned a value 0, any value from 0.75 to 1.5 will be assigned value of "1", and so on.

So using Table 3.11 will be discretized as follows (given in Table 3.12).

3.5 Miscellaneous Concepts

Now we will discuss some miscellaneous concepts taken from [3]. For a decision system A = (U, C ∪ {d}), the cardinality of the image d(U) is called rank of decision attribute "d" and is denoted by r(d). For example, for decision system in Table 3.2, the rank of decision class is "2".

A decision system can also be partitioned on the basis of decision attribute. $\text{CLASS}_A(d) = \left\{X_A^1, X_A^2, \ldots, X_A^{r(d)}\right\}$ is called classification of objects in A determined by d, where X_A^i is called *i*th-decision class.

If $\left\{X_A^1, X_A^2, \ldots, X_A^{r(d)}\right\}$ are decision classes of decision system A, then $\{\underline{C}_{x1} \cup \underline{C}_{x2} \cdots \underline{C}_{r(d)}\}$ is called positive region.

For decision system given in Table 3.2, there are two decision classes, i.e. {1,0}. The partitioning of the universe for this table using decision attribute is $U = \{X^1 \cup X^2\}$ where $X^1 = \{u1, u4, u6, u7\}$ and $X^0 = \{u2, u3, u5\}$.

For a decision system A = (U, C ∪ {d}), the decision system is consistent if and only $\text{POS}_A(d) = U$.

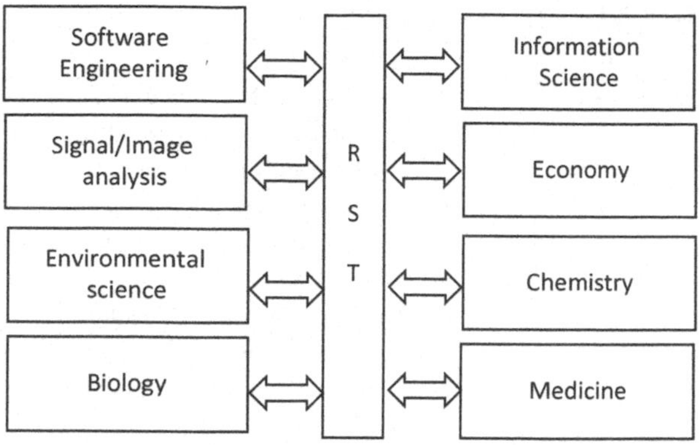

Fig. 3.9 Graphical representation of few RST application

3.6 Applications of RST

Right from its inception, it has been used in various domains for data analysis including economy and finance [76], medical diagnosis [77], medical imaging [78], banking [79], data mining [80], etc. Figure 3.9 shows a graphical representation of different applications of RST.

Here we present some of the representative applications of RST in different domains. Table 3.13 shows sample applications of RST in different domains.

3.7 Summary

This chapter has in-depth discussion of the preliminary concepts of rough set theory. To start RST, it was necessary to have some basic concepts of classical set theory. So, for this purpose, some basic concepts of classical set theory were also discussed. For RST each and every concept was complemented with example to help reader grasp the concept. We have also provided details of some miscellaneous topics including discretization process and some other definitions.

A thorough literature review regarding use of RST in various domains along with its exact application has also been made part of the chapter. Special efforts were made to keep the presentation method simple and basic, so that it could help the readers understand RST that has always been a tough task to learn due to its mathematical nature. In the upcoming chapter, we will discuss some advance topics in RST.

Table 3.13 Applications of RST in different domains[3]

Domain	Application	References
Medicine	Treatment of duodenal ulcer by HSV	[4–8]
	Analysis of data from peritoneal lavage in acute pancreatitis	[9–10]
	Knowledge acquisition in nursing	[11]
	Diagnosis of pneumonia patients	[12]
	Medical databases (e.g. headache, meningitis, CVD) analysis	[13]
	Image analysis for medical applications	[14]
	Surgical wound infection	[15]
	Classification of histological pictures	[16]
	Preterm birth prediction	[17]
	Verification of indications for treatment of urinary stones by extracorporeal shock wave lithotripsy (ESWL)	[18]
	Analysis of factors affecting the differential diagnosis between viral and bacterial meningitis	[19–20]
	Developing an emergency room for diagnostic check list A case study of appendicitis	[21]
	Analysis of medical experience with urolithiasis patients treated by extracorporeal shock wave lithotripsy	[22]
	Diagnosing in progressive encephalopathy	[23–25]
	Rough set-based filtration of sound applicable to hearing prostheses	[26]
	Discovery of attribute dependencies in experience with multiple injured patients	[27]
	Modelling cardiac patient set residuals	[28]
	Multistage analysis of therapeutic experience with acute pancreatitis	[29]
	Breast cancer detection using electropotentials	[30]
	Analysis of medical data of patients with suspected acute appendicitis	[31]
	Attribute reduction in a database for hepatic diseases	[32]
	EEG signal analysis	[33]
Economics finance and business	Evaluation of bankruptcy risk	[34–36]
	Company evaluation	[37]
	Customer behaviour patterns	[38]
	Response modelling in database marketing	[39]
	Analysis of factors affecting stock price fluctuation	[40]
	Discovery of strong predictive rules for stock market	[41]
	Purchase prediction in database marketing	[42]
	Modelling customer retention	[43]
	Temporal patterns	[44]
	Analysis of business databases	[45]
	Rupture prediction in a highly automated production system	[46]

(continued)

Table 3.13 (continued)

Domain	Application	References
Environmental cases	Analysis of a large multispecies toxicity database	[47]
	Drawing premonitory factors for earthquakes by emphasizing gas geochemistry	[48]
	Control conditions on a polder	[49]
	Global warming: influence of different variables on the earth global temperature	[50]
	Global temperature stability	[51]
	Programming water supply systems	[52–53]
	Predicting water demands in Regina	[54]
	Prediction of slope-failure danger level from cases	[55]
Signal and image analysis	Noise and distortion reduction in digital audio signal	[56–57]
	Filtration and coding of audio	[58]
	Recognition of musical sounds	[59]
	Detection and interpolation of impulsive distortions in old audio recordings	[60]
	Subjective assessment of sound quality	[61]
	Classification of musical timbres and phrases	[62]
	Image analysis	[14]
	Converting a continuous tone image into a half-tone image using error diffusion and rough set methods	[63]
	Voice recognition	[26]
	Handwritten digit recognition	[64]
Software Engineering	Qualitative analysis of software engineering data	[65]
	Assessing software quality	[66]
	Software deployability	[67]
	Knowledge discovery form software engineering data	[68]
Information sciences	Information retrieval	[69]
	Analysis and synthesis of concurrent systems	[70]
	Integration RDMS and data mining tools using rough sets	[71]
	Rough set model of relational databases	[72]
	Cooperative knowledge-based systems	[73]
Molecular biology	Discovery of functional components of proteins from amino acid sequences	[74]
Chemistry: pharmacy	Analysis of relationship between structure and activity of substances	[75]

Bibliography

1. http://www.rapidtables.com/math/symbols/Basic_Math_Symbols.htm. Accessed 30 Mar 2017
2. Pawlak Z (1991) Rough Sets: Theoretical Aspects of Reasoning About Data. Kluwer Academic, Dordrecht
3. Pal SK, Skowron A (1999) Rough-fuzzy hybridization: a new trend in decision making. Springer-Verlag, New York

4. Pawlak Z, Słowiński K, Słowiński R (1986) Rough classification of patients after highly selective vagotomy for duodenal ulcer. Int J Man-Mach Stud 24(5):413–433
5. Fibak J et al (1986) Rough sets based decision algorithm for treatment of duodenal ulcer by HSV. Biol Sci 34:227–249
6. Fibak J, Slowinski K, Slowinski R (1986) The application of rough set theory to the verification of indications for treatment of duodenal ulcer by HSV, In: Proceedings of 6th Internat, workshop on expert systems and their applications, Avignon, 1: 587–599
7. Slowinski R, Slowinski K (1989) An expert system for treatment of duodenal ulcer by highly selective vagotomy (in Polish). In: Pamietnik 54. Jubil. Zjazdu Towarzystwa Chirurgow Polskich, Krakow I, pp 223–228
8. Slowinski K (1992) Rough classification of HSV patients. In: Intelligent decision support-handbook of applications and advances of the rough sets theory: 77–94
9. Słowiński K (1994) Rough sets approach to analysis of data of diagnostic peritoneal lavage applied for multiple injuries patients. In: Rough sets, fuzzy sets and knowledge discovery. Springer, London, pp 420–425
10. Slowiński K, Slowiński R, Stefanowski J (1988) Rough sets approach to analysis of data from peritoneal lavage in acute pancreatitis. Med Inform 13(3):143–159
11. Grzymala-Busse JW (1998) Applications of the rule induction system LERS. In: Rough sets in knowledge discovery 1, pp 366–375
12. Paterson GI (1994) Rough classification of pneumonia patients using a clinical database. In: Rough sets, fuzzy sets and knowledge discovery. Springer, London, pp 412–419
13. Tsumoto S, Tanaka H (1995) PRIMEROSE: probabilistic rule induction method based on rough sets and resampling methods. Comput Intell 11(2):389–405
14. Jelonek J et al (1994) Neural networks and rough sets—comparison and combination for classification of histological pictures. In: Rough sets, fuzzy sets and knowledge discovery. Springer, London, pp 426–433
15. Kandulski M, Marciniec J, Tukałło K (1992) Surgical wound infection—conducive factors and their mutual dependencies. In: Intelligent decision support. Springer, Dordrecht, pp 95–110
16. Jelonek J et al (1994) Neural networks and rough sets—comparison and combination for classification of histological pictures. In: Rough sets, fuzzy sets and knowledge discovery. Springer, London, pp 426–433
17. Grzymala-Busse JW, and LK Goodwin (1996) A comparison of less specific versus more specific rules for preterm birth prediction. In: Proceedings of the first online workshop on soft computing WSC1 on the Internet, Japan
18. Slowinski K et al (1995) Rough set approach to the verification of indications for treatment of urinary stones by extracorporeal shock wave lithotripsy (ESWL). In: Soft computing, society for computer simulation. San Diego, California, pp 142–145
19. Tsumoto S, Ziarko W (1996) The application of rough sets-based data mining technique to differential diagnosis of meningoenchephalitis. International symposium on methodologies for intelligent systems. Springer, Berlin/Heidelberg
20. Ziarko W (1998) Rough sets as a methodology for data mining. In: Rough sets in knowledge discovery 1. Physica-Verlag, Heidelberg, pp 554–576
21. Rubin S, Michalowski W, Slowinski R (1996) Developing an emergency room diagnostic check list using rough sets-a case study of appendicitis. In: Simulation in the medical sciences, pp 19–24
22. Slowinski K, Stefanowski J (1996) On limitations of using rough set approach to analyse non-trivial medical information systems
23. Paszek P, Wakulicz Deja A (1996) Optimalization diagnose in progressive encephalopathy applying the rough set theory. Zimmermann 557(1):192–196
24. Wakulicz-Deja A, Boryczka M, Paszek P (1998) Discretization of continuous attributes on decision system in mitochondrial encephalomyopathies. In: International conference on rough sets and current trends in computing. Springer, Berlin/Heidelberg

25. Wakulicz-Deja A, Paszek P (1997) Diagnose progressive encephalopathy applying the rough set theory. Int J Med Inform 46(2):119–127
26. Czyzewski A (1998) Speaker-independent recognition of isolated words using rough sets. Inform Sci 104(1-2):3–14
27. Stefanowski J, Slowiński K (1997) Rough set theory and rule induction techniques for discovery of attribute dependencies in medical information systems. In: European symposium on principles of data mining and knowledge discovery. Springer, Berlin/Heidelberg
28. Ohrn A et al (1997) Modelling cardiac patient set residuals using rough sets. In: Proceedings of the AMIA annual fall symposium. American Medical Informatics Association
29. Słowiński K, Stefanowski J (1998) Multistage rough set analysis of therapeutic experience with acute pancreatitis. In: Rough sets in knowledge discovery 2. Physica-Verlag, Heidelberg, pp 272–294
30. Swiniarski RW (1998) Rough sets and bayesian methods applied to cancer detection. In: International conference on rough sets and current trends in computing. Springer, Berlin/Heidelberg
31. Carlin US, Komorowski J, Øhrn A (1998) Rough set analysis of patients with suspected acute appendicitis. In: Traitement d'information et gestion d'incertitudes dans les systèmes à base de connaissances. Conférence internationale
32. Tanaka H, Maeda Y (1998) Reduction methods for medical data. In: Rough sets in knowledge discovery 2. Physica-Verlag, Heidelberg, pp 295–306
33. Wojdyłło P (1998) Wavelets, rough sets and artificial neural networks in EEG analysis. In: International conference on rough sets and current trends in computing. Springer, Berlin/Heidelberg
34. Slowinski R, Zopounidis C (1995) Application of the rough set approach to evaluation of bankruptcy risk. Intell Syst Account Financ Manag 4(1):27–41
35. Slowinski R, Zopounidis C (1994) Rough-set sorting of firms according to bankruptcy risk. In: Applying multiple criteria aid for decision to environmental management. Springer, Dordrecht, pp 339–357
36. Greco S, Matarazzo B, Slowinski R (1998) A new rough set approach to evaluation of bankruptcy risk. In: Operational tools in the management of financial risks. Springer, pp 121–136
37. Mrózek A, Skabek K (1998) Rough sets in economic applications. In: Rough sets in knowledge discovery 2. Physica-Verlag HD, Heidelberg, pp 238–271
38. Piasta Z, Lenarcik A (1998) Learning rough classifiers from large databases with missing values. In: Rough sets in knowledge discovery 1, pp 483–499
39. Van den Poel D (1998) Rough sets for database marketing. In: Rough sets in knowledge discovery 2. Physica-Verlag HD, Heidelberg/New York, pp 324–335
40. Golan RH, Ziarko W (1995) A methodology for stock market analysis utilizing rough set theory. In: Computational intelligence for financial engineering, proceedings of the IEEE/IAFE 1995. IEEE, 1995
41. Ziarko W, Golan R, Edwards D (1993) An application of datalogic/R knowledge discovery tool to identify strong predictive rules in stock market data. In: Proceedings of AAAI workshop on knowledge discovery in databases, Washington, DC
42. Van den Poel D, Piasta Z (1998) Purchase prediction in database marketing with the ProbRough system. In: International conference on rough sets and current trends in computing. Springer, Berlin/Heidelberg
43. Kowalczyk, AE Eiben TJ Euverman W, Slisser F (1999) Modelling customer retention with statistical techniques, rough data models, and genetic programming. In: Rough fuzzy hybridization: a new trend in decision-making
44. Kowalczyk W (1996) Analyzing temporal patterns with rough sets. Zimmermann 557:139
45. Kowalczyk W, Piasta Z (1998) Rough-set inspired approach to knowledge discovery in business databases. In: Pacific-asia conference on knowledge discovery and data mining. Springer, Berlin/Heidelberg

46. Swiniarski R et al (1997) Feature selection using rough sets and hidden layer expansion for rupture prediction in a highly automated production process. Syst Sci Wroclaw 23:53–60
47. Keiser K, Szladow A, Ziarko W (1992) Rough sets theory applied to a large multispecies toxicity database. In: Proceedings of the fifth international workshop on QSAR in environmental toxicology, Duluth
48. Teghem J, Charlet J-M (1992) Use of "rough sets" method to draw premonitory factors for earthquakes by emphasing gas geochemistry: the case of a low seismic activity context, in Belgium. In: Intelligent Decision Support. Springer, Dordrecht, pp 165–179
49. Reinhard A et al (1992) An application of rough set theory in the control conditions on a polder. Slowinski 428:331
50. la Busse, JW Grzyma, and Gunn JD (1995) Global temperature analysis based on the rule induction system LERS. In: Proceedings of the fourth international workshop on intelligent information systems, Augustow, Poland, June, vol 5, No 9
51. Gunn JD, Grzymala-Busse JW (1994) Global temperature stability by rule induction: An interdisciplinary bridge. Hum Ecol 22(1):59–81
52. Greco S, Matarazzo B, Słowiński R (1998) Rough approximation of a preference relation in a pairwise comparison table. In: Rough sets in knowledge discovery 2. Physica-Verlag HD, Heidelberg, pp 13–36
53. Roy B, Slowinski R, Treichel W (1992) Multicriteria Programming Of Water Supply Systems For Rural Areas1. J Am Water Resour Assoc 28:13–31
54. An A et al. (1995) Discovering rules from data for water demand prediction. In: Proceedings of the workshop on machine learning in engineering IJCAI. Vol. 95
55. Furuta H, Hirokane M, Mikumo Y (1998) Extraction method based on rough set theory of rule-type knowledge from diagnostic cases of slope-failure danger levels. In: Rough sets in knowledge discovery 2. Physica-Verlag, Heidelberg, pp 178–192
56. Czyzewski A (1996) Mining knowledge in noisy audio data. KDD
57. Czyżewski A (1998) Soft processing of audio signals. In: Rough sets in knowledge discovery 2. Physica-Verlag, Heidelberg, pp 147–165
58. Czyzewski A and Krolikowski R (1997) New methods of intelligent filtration and coding of audio. In: Audio Engineering Society Convention 102. Audio Engineering Society
59. Kostek B (1998) Soft computing-based recognition of musical sounds. In: Rough sets in knowledge discovery 2. Physica-Verlag, Heidelberg, pp 193–213
60. Czyzewski A (1995) Some methods for detection and interpoleation of impulsive distortions in old audio recordings. In: Applications of signal processing to audio and acoustics, IEEE ASSP Workshop on. IEEE, 1995
61. Kostek B (1998) Soft set approach to the subjective assessment of sound quality. In: Fuzzy systems proceedings, 1998. IEEE world congress on computational intelligence., The 1998 I. E. international conference on. Vol. 1. IEEE
62. Kostek B and Szczerba M (1996) Parametric representation of musical phrases. In: Audio engineering society convention 101. Audio Engineering Society
63. Zeng H, Swiniarski R (1998) A new halftoning method based on error diffusion with rough set filtering. In: Rough sets in knowledge discovery 2. Physica-Verlag, Heidelberg, pp 336–342
64. Bazan JG et al (1998) Synthesis of decision rules for object classification. In: Incomplete information: rough set analysis. Physica-Verlag, Heidelberg, pp 23–57
65. Ruhe G (1996) Qualitative analysis of software engineering data using rough sets. In: Tsumoto, Kobayashi, Yokomori, Tanaka, and Nakamura 484, p 292
66. Peters JF, Ramanna S (1999) A rough sets approach to assessing software quality: Concepts and rough Petri net models. In: Rough-fuzzy hybridization: new trends in decision making. Springer, Berlin, pp 349–380
67. Peters JF and S Ramanna (1998) Software deployability decision system framework: a rough set approach. In: Traitement d'information et gestion d'incertitudes dans les systèmes à base de connaissances. Conférence internationale

68. Ruhe G (1997) Knowledge discovery from software engineering data: rough set analysis and its interaction with goal-oriented measurement. In: European symposium on principles of data mining and knowledge discovery. Springer, Berlin/Heidelberg
69. Srinivasan P (1989) Intelligent information retrieval using rough set approximations. Inf Process Manag 25(4):347–361
70. Skowron A and Suraj Z. (1993) Rough sets and concurrency. Bull Acad Pol Sci. Technical sciences 41.3: pp 237–254
71. Nguyen SH et al. (1996) Knowledge discovery by rough set methods. In: Proceedings of the international conference on information systems analysis and synthesis ISAS, vol 96
72. Beaubouef T, Petry FE (1994) A rough set model for relational databases. In: Rough sets, fuzzy sets and knowledge discovery. Springer, London, pp 100–107
73. Ras ZW (1996) Cooperative knowledge-based systems. Intell Autom Soft Comput 2 (2):193–201
74. Tsumoto S and H Tanaka (1995) Automated discovery of functional components of proteins from amino-acid sequences based on rough sets and change of representation. KDD
75. Krysiński J (1990) Rough sets approach to the analysis of the structure-activity relationship of quaternary imidazolium compounds. Arztl Forsch 40(7):795–799
76. Podsiadło M, Rybiński H (2014) Rough sets in economy and finance. In: Transactions on rough sets XVII. Springer, Berlin/Heidelberg, pp 109–173
77. Prasad V, Srinivasa Rao T, Surendra Prasad Babu M (2016) Thyroid disease diagnosis via hybrid architecture composing rough data sets theory and machine learning algorithms. Soft Comput 20(3):1179–1189
78. Xie C-H, Liu Y-J, Chang J-Y (2015) Medical image segmentation using rough set and local polynomial regression. Multimedia Tools Appl 74(6):1885–1914
79. Montazer GA, ArabYarmohammadi S (2015) Detection of phishing attacks in Iranian e-banking using a fuzzy–rough hybrid system. Appl Soft Comput 35:482–492
80. Maciá-Pérez F et al (2015) Algorithm for the detection of outliers based on the theory of rough sets. Decis Support Syst 75:63–75

Chapter 4
Advance Concepts in RST

In the last chapter, we discussed some basic concepts of rough set theory. In this chapter we will present advance concepts in RST including some improved definitions, their examples and hybridization with fuzzy set theory.

4.1 Fuzzy Set Theory

Later in this chapter, we will discuss few concepts related to hybridization of RST with fuzzy set theory, but we need to first explain some basic concepts of fuzzy set theory as discussed below.

4.1.1 Fuzzy Set

If X is a universe of disclosure and x is a particular element of X, then a fuzzy set F defined on X will be a collection of ordered pairs:

$$A = \{(x, \mu_{\tilde{A}}(x)), x \in X\} \tag{4.1}$$

where each pair $(x, \mu_F(x))$ is called a singleton. In the case of crisp set theory, we don't use $\mu_F(x)$, because for all the elements which are present, their degree would be one and for non-present elements, the degree would be zero. Consider the following set U:

M.S. Raza, U. Qamar, *Understanding and Using Rough Set Based Feature Selection: Concepts, Techniques and Applications*,
DOI 10.1007/978-981-10-4965-1_4

$$U = \{0, 1, 2, 3, 4, 5, 6, 7, 8, 9\}$$
$$\underset{\sim}{A} = \{1, 3, 5, 7, 9\}$$
$$\widetilde{A} = \{(0,0), (1,1), (2,0)(3,1), (4,0), (5,1), (6,0), (7,1), (8,0), (9,1)\}$$

Here note that we have shown ten ordered pairs, one for each element. Each ordered pair shows the element and its degree of presence in the set. For example, the first pair (0,1) shows that the degree of presence of element "0" is "0", i.e. it is present in the set A; on the other hand, the pair (1,1) shows that element "1" is present in "*A*". It should be noted that the degree of an element has the range from zero (0) to one (1). For example, consider the reference set A of intelligent students where "intelligent" is a fuzzy term:

$$A = \{S1, S2, S3, S4, S5\}$$

Now it should be noted that each person has his/her own intelligence level. It is not that a student will have zero intelligence level while the other will have a level of 100%. So, a fuzzy set may take the form

$$\widetilde{A} = \{(S1, 0.1), (S2, 0.4), (S3, 0.5)(S4, 0.5), (S5, 1)\}$$

The above set shows the degree or strength; a student belongs to Set A. Student $S2$ is more intelligent than $S1$; similarly $S5$ is more smart than $S4$. Table 4.1 shows the difference between fuzzy set and concrete set. Figures 4.1 and 4.2 show the graphical representation of both.

Crisp set shows that only Saturday and Sunday will be part of the set of weekend days, while in fuzzy set "Friday" may also have a partial membership; this is because we normally begin to feel the weekend on a Friday evening.

Table 4.1 Difference between fuzzy set and concrete set

Fuzzy	Classic
Do not restrict some member to be fully part of a set	Restrict an object to either be a member of a set or not
Allow partial membership	Don't allow partial membership
If we consider a fuzzy set of weekends, Friday may be partial member of set	Only Saturday and Sunday will be members

Fig. 4.1 Fuzzy vs crisp set

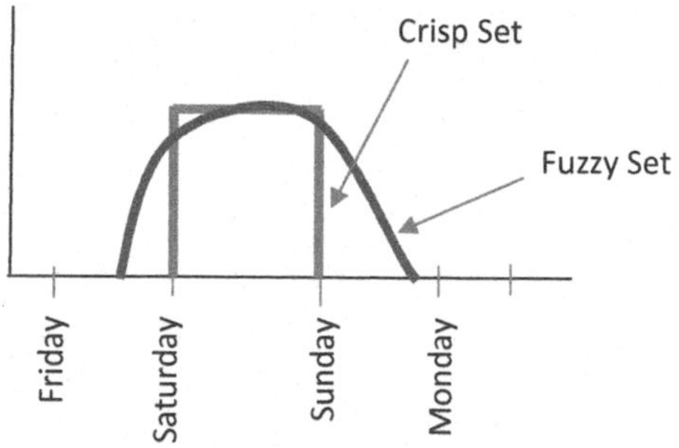

Fig. 4.2 Fuzzy vs crisp set

4.1.2 *Fuzzy Sets and Partial Truth*

As discussed earlier, this universe is not crisp; we come across many situations in real-life scenarios where we find the partial answers, e.g. if you ask about how much work you have done, you will probably get an answer like almost or only a little bit left. Now the question is that what is meant by almost either it means all of the work or 60%, 70% or how much. So the fuzzy logic gives us the ability to answer a question with partial degree of certainty in formation. Fuzzy logic thus generalizes the yes/no or 1/0 logic, i.e. in fuzzy logic, the answer may be 0.72 or 0.5 instead of exact 1 or 0. So in fuzzy logic, truth has its degree, a value indicating the amount of truth in a statement, e.g. consider the following questions and their answers:

Q: Are you in seventh class?
Answer: No (crisp)
Q: Are you going to New York?
Answer: Yes (crisp)
Question: How much do you like to go for camping on weekends?
Answer: Not too much (fuzzy)
Question: How much excited are you on weekends?
Answer: Very much (fuzzy)

If you take note of the last two questions, the answer is somewhere between "yes" and "no" in contrast with the first two questions.

4.1.3 *Membership Function*

The degree of truth that we have talked about is determined by membership function. A membership function is a graph that determines the elements of the input space to the degree of truth (membership) between 0 and 1. It can range from a simple linear function to a complex polynomial one.

The following are some characteristics of a membership function:

Fig. 4.3 Fuzzy vs crisp set

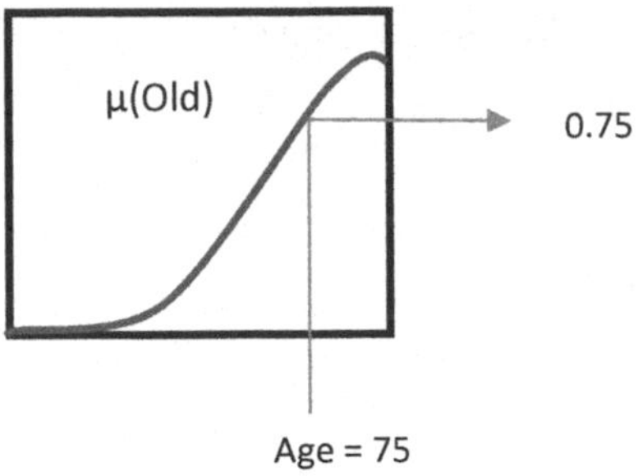

- It is represented by the symbol μ.
- It takes elements from the input space and returns their degree of membership.
- The degree of membership ranges from zero to one, where zero represents absolute falseness and one represents absolute truth.

Figure 4.3 represents a membership function μ(Old) that determines the degree of membership (truth) of a person's age to the set of old people.

So, if a person's age is 75 years, the membership function μ(Old) returns to $\mu = 0.75$ which means that the person is quite old. Note that 0.75 just provides information about the tendency of this person to belong to the set of old persons.

4.1.4 *Fuzzy Operators*

Before we move on to fuzzy operators, let's take a look at conventional logical AND, OR and NOT operators.

AND Operator

A	B	A AND B
0	0	0
0	1	0
1	0	0
1	1	1

OR Operator

A	B	A OR B
0	0	0
0	1	1
1	0	1
1	1	1

NOT Operator

A	NOT A
0	1
1	0

Conventional logical operators

The table provides the Boolean input and corresponding crisp values after applying the AND, OR and NOT operators. Fuzzy operators on the other hand can act on real numbers as well. In fuzzy set theory, we use Min (minimum), Max (maximum) and 1-A for AND, OR and NOT operators, respectively. The terms intersection, union and complement are also used.

4.1.4.1 Union

Let A and B be two fuzzy sets. Ifμ(A) and μ(B) are two membership functions on universe X, then the union of μ(A) and μ(B) can be defined by fuzzy union operator as follows:

$$\mu_{A \cup B}(X) = \text{MAX}(\mu(\text{A}), \mu(\text{B})) \tag{4.2}$$

Table 4.2 explains the union operator for two fuzzy sets A and B.

4.1.4.2 Intersection

Let A and B be two fuzzy sets. Ifμ(A) and μ(B) are two membership functions on universe X, then the union of μ(A) and μ(B) can be defined by fuzzy intersection operator as follows:

$$\mu_{A \cap B}(X) = \text{MIN}(\mu(\text{A}), \mu(\text{B})) \tag{4.3}$$

Table 4.3 explains the intersection operator for two fuzzy sets A and B.

4.1.4.3 Complement

Let A be a fuzzy set, and if μ(A) is a membership function on universe X, then the complement operator is defined as

$$\mu_{A^c}(X) = 1 - \mu(\text{A}) \tag{4.4}$$

Table 4.4 explains the complement operator for the fuzzy set A.
Figure 4.4 shows all the three operators.

Table 4.2 Union operator

μ(A)	μ(B)	$\mu_{A \cup B}(X) = \text{MAX}(\mu(\text{A}), \mu(\text{B}))$
0.5	0.2	0.5
0.1	0.05	0.1
0.0	0.0	0.0
0.3	0.35	0.35

Table 4.3 Intersection operator

μ(A)	μ(B)	$\mu_{A \cup B}(X) = \text{MIN}(\mu(\text{A}), \mu(\text{B}))$
0.5	0.2	0.2
0.1	0.05	0.05
0.0	0.0	0.0
0.3	0.35	0.3

Table 4.4 Complement operator

μ(A)	$\mu_{A^c}(X) = 1 - \mu(A)$
0.5	0.5
0.1	0.9
0.0	1.0
0.3	0.7

Fig. 4.4 Fuzzy operators

4.1.5 Fuzzy Set Representation

A fuzzy set can be represented by two ways. The first one is by using a triangular graph, the same as given in the figure above where the peak represents the mean value. This is based on the assumption that most of the population lies at average or mean value whereas exceptional cases are represented by the far edges on the slope of the triangle. The other and most convenient way to represent fuzzy sets is to represent a set in the form of value and membership pair. For example, to represent a fuzzy set "Tall" about the height of people, we can use the following notation:

$$Tall = \{0/4, 0/4.5, 0/5, 0.25/5.5, 0.5/6, 0.75/6.5, 1/7\}$$

Here the numerator represents the membership value and the denominator represents the actual height. So a person with the height up to 4.5 inches will not be considered tall, whereas a person with the height 7 inches will definitely be considered tall.

4.1.6 Fuzzy Rules

Normally rules are implemented in the form of If-Else statements, e.g. from Table 4.5, we can have two rules as follows:

$$If\ X = A\ THEN\ Y = 0$$

This was a simple rule; we can have more complicated rules as well, e.g. from the above table, we can also derive that

Table 4.5 A raw dataset with two variables

X	Y
A	0
B	1
A	0
C	1

$$IF\ (X = B)\ OR\ (X = C)\ THEN\ Y = 1$$

Here X and Y are variables and {B,C,0,1} are fuzzy distributions or sets. A more real-life rule may be

If slope is steep, THEN energy consumption is high.

Here the variable slope, in the fuzzy system, may have crisp values, e.g. some number from 1 to 5. A rule has two parts called antecedents and consequent.

Antecedents are the conditions that need to be fulfilled to get the result, e.g. if the weather is good, then we can go out for camping. As discussed above, antecedents may have multiple parts concatenated using logical operators, in which case all parts are resolved to some single number.

Consequent is the resultant part, i.e. the portion after the word "THEN". Consequent may also have multiple parts, e.g. in the following rule:

If slope is steep, THEN energy consumption is high and speed is slow.

We have two consequent parts. Antecedents affect consequents. We use implication function to modify the output fuzzy set to the degree specified by antecedents, e.g. for the rule

If slope is steep OR road is rough, THEN energy consumption is high.

The fuzzy system works as given in Fig. 4.5. The process comprises three steps:

Step 1: Fuzzify the Inputs

All the antecedents are resolved to a single value, the degree of membership between 0 and 1. For example, in the above-mentioned rule, the slope was given a value 60% which fuzzifies to 0.6 by the fuzzy membership function in the fuzzy set of high slope. Similarly roughness of road was given a value 50% which resolved to 0.5 in the fuzzy set of roughness.

Step 2: Apply Fuzzy Operators

If the rule comprises multiple antecedents, then fuzzy operators are applied to solve the antecedent values. For example, in our example, there are two antecedents, i.e. "slope" and "road". There is an OR operator between them, so max(0.6, 0.5) results in 0.6 called degree of support for rule.

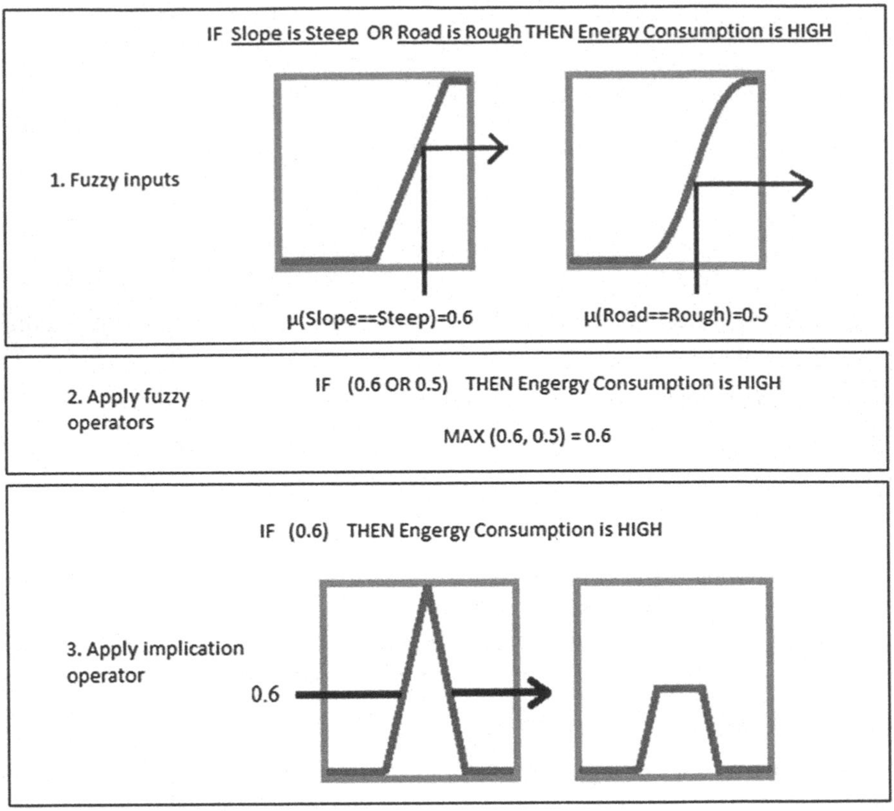

Fig. 4.5 How fuzzy system works

Step 3: Apply Implication Method

Apply implication method and use the degree of support to shape the output fuzzy set. The consequent of a fuzzy rule assigns an entire fuzzy set to the output. This fuzzy set is represented by a membership function that is chosen to indicate the qualities of the consequent. Then the output fuzzy set is truncated according to the implication method.

4.2 Fuzzy-Rough Set Hybridization

Both fuzzy and rough set theories are important components of computations. Rough set theory models uncertainty whereas fuzzy sets model vagueness. Researchers have already exploded a number of ways in which both theories can interact with each other [4]. Here we will first see some of the applications of fuzzy-rough hybridization in supervised learning, information retrieval and feature selection.

4.2.1 Supervised Learning and Information Retrieval

In [5] authors have enhanced the classification accuracy of K-nearest neighbours algorithm using fuzzy-rough uncertainty, however still maintaining the simplicity and nonparametric characteristics. In proposed solution, we do not need to know the value of "K" as in case of conventional one. Furthermore, the class confidence values, calculated using fuzzy-rough ownership values, do not sum up to one which helps algorithm distinguish between equal evidence and ignorance and thus make the semantics of the class confidence values richer.

Natural languages contain a lot of uncertainty and vagueness, so fuzzy-rough hybridization can better help us solve natural language-related problems. In [6] authors have used fuzzy-rough framework for query refinement. They define a thesaurus using upper approximation where the query is approximated from both upper and lower side. The upper approximation was related to query explosion, whereas the lower approximation was found to be too strict resulting in empty query. So authors propose the use of lower approximation of the upper approximation (different from upper approximation) in case when thesaurus is not transitive.

4.2.2 Feature Selection

So far RST has been successfully applied for feature selection. However, approaches have been proposed to use fuzzy-rough hybridization. In [7] authors have presented a fuzzy-rough set-based ant colony optimization algorithm and implemented it using fuzzy-rough hybridized approach to find the optimal feature subset.

The data in real world normally exists in hybrid format, so an effective technique for reduction of such data is also desirable. In [8] the author proposes an information measure to compute the discernibility power of a fuzzy or crisp equivalence relation, and by using this measure, the significance of different attributes was measured. Finally authors have also proposed two reduction algorithms for both supervised and unsupervised datasets.

Table 4.6 presents few of the feature selection techniques based on fuzzy-rough sets.

4.2.3 Rough Fuzzy Set

Dubios et al. in [9] defined the concept of rough fuzzy sets to model the situation where the knowledge base contains crisp concepts and output classes have poorly defined boundaries. It is rough set that is approximately deduced from fuzzy set in a crisp and approximate space. Output class is fuzzy in rough fuzzy set. For a rough fuzzy set FX, the lower and upper approximations are defined as follows:

Table 4.6 Fuzzy-rough feature selection approaches

Fuzzy-rough approach	Description
New approaches to fuzzy-rough feature selection [9]	Research proposes new approaches to fuzzy-rough FS based on fuzzy similarity relations. A fuzzy extension to crisp discernibility matrices is presented as well
Different classes' ratio fuzzy-rough set-based robust feature selection [10]	Research proposes an effective robust fuzzy-rough set model, called different classes' ratio fuzzy-rough set (DC_ratio FRS) model, to reduce the influence of noisy samples on the computation of the lower and upper approximations and recognize the noisy samples directly
Large-scale multimodality attribute reduction with multi-kernel fuzzy-rough sets [11]	Research defines a combination of kernels based on set theory to extract fuzzy similarity for fuzzy classification with multimodality attributes. Then, a model of multi-kernel fuzzy-rough sets is constructed. Finally, the paper presents an attribute reduction algorithm for large-scale multimodality fuzzy classification based on the proposed model
Towards scalable fuzzy-rough feature selection [12]	The paper proposes two different novel ways to address this problem using a neighbourhood approximation step and attribute grouping in order to alleviate the processing overhead and reduce complexity
Fuzzy-rough feature selection accelerator [13]	The paper proposes an accelerator, called forward approximation, which combines sample reduction and dimensionality reduction together. The strategy can be used to accelerate a heuristic process of fuzzy-rough feature selection. Based on the proposed accelerator, an improved algorithm is also designed
Semi-supervised fuzzy-rough feature selection [14]	The paper proposes a novel approach for semi-supervised fuzzy-rough feature selection where the object labels in the data may only be partially present. The approach also has the appealing property that any generated subsets are also valid (super)reducts when the whole dataset is labelled
Feature selection algorithm using fuzzy-rough sets for predicting cervical cancer risks [15]	Research uses fuzzy-rough set method to analyse the demographic dataset and identify the risk of cervical cancer. This method integrates entropy, information gain (IG) and fuzzy-rough sets for identifying the risk of cervical cancer earlier. Risk factors are identified by IG. Rules are extracted by fuzzy-rough sets

Suppose FS = (U, A, V, f) is knowledge representation system. If P ⊆ A and FX ⊆ U are a fuzzy set, then the lower approximation $\underline{apr}_P(FX)$ and upper approximation $\overline{apr}_P(FX)$ of FS about FX are defined as

$$\underline{apr}_P(FX) = \inf\{x \in I\ (x) : \mu_{FX}(x)\} \quad (4.5)$$

$$\overline{apr}_P(FX) = \sup\{\ x \in \mathrm{I}\ (\mathrm{x}) : \mu_{FX}(x)\} \quad (4.6)$$

where I is equivalence relation on U and $\mu_{FX}(x)$ is the degree of membership of x to FX. The lower approximation defines the degree to which an object definitely belongs to fuzzy set FX whereas upper approximation defines the degree of membership to which an object x possibly belongs to FX. So if for the values 0 or 1, these definitions are equal to those in conventional rough set theory.

If U = {X1, X2, X3, X4, X5} is a set of five employees and comprises two equivalence classes U/I = {{X1, X3, X5}{X2, X4}}. Suppose a fuzzy set FX represents the concept "competence" and membership function $\mu_{FX}(X) = \left\{\frac{X1}{0.5}, \frac{X2}{0.4}, \frac{X3}{0.1}, \frac{X4}{0.8}, \frac{X5}{0.7}\right\}$, the upper and lower approximation of the rough fuzzy set will be

$$\underline{apr}_P(FX) = \left\{\frac{X1}{0.1}, \frac{X2}{0.4}, \frac{X3}{0.1}, \frac{X4}{0.4}, \frac{X5}{0.1}\right\}$$

$$\overline{apr}_P(FX) = \left\{\frac{X1}{0.7}, \frac{X2}{0.8}, \frac{X3}{0.7}, \frac{X4}{0.8}, \frac{X5}{0.7}\right\}$$

4.2.4 Fuzzy-Rough Set

Fuzzy-rough sets are extensions of rough fuzzy set. According to Dubios in [9], fuzzy-rough set can be defined as follows:

The pair $\langle \underline{P}X, \bar{P}X \rangle$ is called fuzzy-rough set where $\underline{P}X$ defines lower approximation and $\bar{P}X$ is called upper approximation as follows:

$$\mu_{\underline{P}X}(F_i) = \inf \max\{1 - \mu_{F_i}(x), \mu_X(x)\} \forall \mathrm{I} \quad (4.7)$$

$$\mu_{\bar{P}X}(F_i) = \inf \max\{\mu_{F_i}(x), \mu_X(x)\} \forall \mathrm{I} \quad (4.8)$$

Here:

F_i = *equivalence class*
X = *fuzzy concept that requires approximation*

Obviously, when the equivalence relation is clear, a fuzzy-rough set will degrade into a rough fuzzy set. When all the fuzzy equivalence relations are clear, it will further degrade into a classical rough set. So, the crisp lower approximation can be characterized by the following function [10]:

$$\mu_{\underline{P}X}(x) = \begin{cases} \mathbf{1}, & x \in F, F \subseteq X \\ \mathbf{0}, & otherwise \end{cases} \tag{4.9}$$

This means that if an object X belongs to an equivalence class which is subset of X, the object will belong to lower approximation of X.

4.3 Dependency Classes

In conventional RST, calculating the dependency of an attribute "D" on "C" requires scanning of the dataset and calculating the positive region which is a time-consuming job. In [1] authors have presented a new concept of dependency classes. They developed an alternate way to calculate dependency comprising of dependency classes.

A dependency class is a heuristic which defines how the dependency measure changes as we scan new records during traversal of the dataset.

They start with first record in dataset and calculate the dependency of decision attribute on conditional attribute based on the derived heuristics. Then after adding each single record, the dependency of a particular attribute is refreshed based on to which decision class the value of that attribute leads to. On the basis of the heuristics used by dependency classes, two types of dependency classes are proposed as follows:

- Incremental dependency classes
- Direct dependency classes

4.3.1 *Incremental Dependency Classes (IDC)*

Incremental dependency classes comprise four rules where each rule defines a class that governs how dependency of decision attribute "D" on "C" changes as we read each new record.

We will explain each incremental dependency class with the help of the example. Consider the following decision system shown in Table 4.7 taken from [2]:

Here:

$C = \{a', b', c', d'\}$

$D = \{D\}$ *and* $|U| = 10$

Table 4.7 Decision system example

U	a′	b′	c′	d′	D
A	M	L	3	M	1
B	M	L	1	H	1
C	L	L	1	M	1
D	L	R	3	M	2
E	M	R	2	M	2
F	L	R	3	L	3
G	H	R	3	L	3
H	H	N	3	L	3
I	H	N	2	H	2
J	H	N	2	H	1

Initially we start with the |U| = 6 where U = {a, b, d, e, f, g}. We calculate the dependency of "D" on all attributes present in C (given at the end of each column in Table 4.8).

Now we define different classes through which the dependency can be calculated, after adding a new record.

4.3.1.1 Existing Boundary Region Class

For an attribute a', if the same value of attribute leads to different decision classes, for example, in Table 4.8, a′(L)->2,3 (i.e. the value "L" leads to decision classes "2" and "3"), then adding a new record with same value of a' decreases the dependency of decision on that attribute. Adding a row in this case will simply decrease the dependency.

For example, in Table 4.8,

$$\gamma(a', D) = \frac{1}{6}$$

After adding a new record, i.e. object "C", the new dataset with new dependency values is shown in Table 4.9.

Before adding new record: a′(L)-> 2,3 (in Table 4.8).

After adding new record: a′(L)-> 1,2,3 (in Table 4.9).

So by adding a new record, the value "L" of attribute "a'", which initially was leading to two decision classes, now leads to three decision classes, so γ(a′,D) becomes

$$\gamma(a', D) = \frac{1}{7}$$

Table 4.8 Decision system example

U	a′	b′	c′	d′	D
A	M	L	3	M	1
B	M	L	1	H	1
D	L	R	3	M	2
E	M	R	2	M	2
F	L	R	3	L	3
G	H	R	3	L	3
	0.16667	**0.3333**	**0.3333**	**0.5**	

Table 4.9 Adding new object "C"

U	a′	b′	c′	d′	D
A	M	L	3	M	1
B	M	L	1	H	1
C	L	L	1	M	1
D	L	R	3	M	2
E	M	R	2	M	2
F	L	R	3	L	3
G	H	R	3	L	3
	0.14286	**0.4286**	**0.4286**	**0.4286**	

4.3.1.2 Positive Region Class

For an attribute a′, if adding a record does not lead to a different decision class for the same value of that attribute, then dependency will increase. For example, in Table 4.8, b′(L)->1. The previous dependency value was 2/6. After adding a new row (object "C" as shown in Table 4.9), b′(L)->1 sustains, i.e. the value "L" of attribute b' uniquely identifies the decision class, so the new dependency will be

$$\gamma(b', D) = \frac{3}{7}$$

4.3.1.3 Initial Positive Region Class

For an attribute a′, if the value appears in the dataset for the first time for that attribute, then dependency increases. For example, adding a new record (object "I") as shown in Table 4.10,

b′(N)->2. Initially the value "N" for b′ attribute was not present. Now adding the record for this value of b′ leads to a new dependency value as follows:

$$\gamma(b, D) = \frac{4}{8}$$

Table 4.10 Adding new object "I"

U	a′	b′	c′	d′	D
A	M	L	3	M	1
B	M	L	1	H	1
C	L	L	1	M	1
D	L	R	3	M	2
E	M	R	2	M	2
I	H	N	2	H	2
F	L	R	3	L	3
G	H	R	3	L	3
	0/8	**0.5**	**0.5**	**0.25**	

4.3.1.4 Boundary Region Class

For an attribute a′, if the same value (which was leading to a unique decision previously) of attribute leads to a different decision, then adding the new record reduces the dependency.

For example, adding a record "H" in Table 4.11, the new dependency $\gamma(b',D)$ will be

$$\gamma(c,D)=\frac{1}{9}$$

which was $\gamma(c,D)=\frac{3}{9}$ before adding record "H". Table 4.12 shows the summary of all decision classes.

4.3.1.5 Mathematical Representation of IDC

Now we provide the mathematical representation of IDC and an example about how to calculate IDC. Mathematically

$$\gamma(Attribute,D)=\frac{1}{N}\sum\nolimits_{k=1}^{N}\gamma'_k \tag{4.10}$$

Where:

$$\gamma'_k=\begin{cases}1, & \text{if } V_{\text{Attribute},k} \text{ leades to a positive region class}\\ 1, & \text{if } V_{\text{Attribute},k} \text{ leades to an initial positive region class}\\ -n, & \text{if } V_{\text{Attribute},k} \text{ leades to a boundary region class}\\ 0, & x \text{ if } V_{\text{Attribute},k} \text{ has already lead to boundary region class}\end{cases}$$

Table 4.13 represents the symbol and its semantics.

Table 4.11 Adding new object "H"

U	a′	b′	c′	d′	D
A	M	L	3	M	1
B	M	L	1	H	1
C	L	L	1	M	1
D	L	R	3	M	2
E	M	R	2	M	2
I	H	N	2	H	2
F	L	R	3	L	3
G	H	R	3	L	3
H	H	L	3	L	3
	0	**1/9**	**0.5**	**0.25**	

Table 4.12 Summary of IDC

Decision class	Definition	Initial attribute value	After adding new record	Effect on dependency
Existing boundary region class	If the same value of attribute leads to different decision classes, it decreases the dependency	a′(L)-> 2,3	a′(L)-> 1,2,3	Decreases
Positive region class	If adding a record does not lead to a different decision class for the same value of that attribute, then dependency will increase	b′(L)->1	b′(L)->1	Increases
Initial positive region class	If the value appears in the dataset for the first time for that attribute, then dependency increases	–	b′(N)->2	Increases
Boundary region class	If the same value (which was leading to unique decision previously) of attribute leads to a different decision, then adding the new record reduces the dependency	b′(L)->1	b′(L)-> 1,3	Decreases

4.3.1.6 Example

The following example shows how IDC calculates dependency. We read each record and identify its dependency class. Based on the class, we decide the factor by which dependency will be added in overall dependency value. We will consider the dataset given in Table 4.7. Using IDC,

$$\gamma(\text{Attribute}, D) = \frac{1}{N}\sum_{k=1}^{N}\gamma'_k$$

Consider the attribute {a'}; we read first three records, i.e. object "A", as it appears for the first time, so it belongs to "initial positive region" class; thus we will add "1" to the overall dependency value. Reading objects "B" and "C" lead to

Table 4.13 Symbols and their semantics

Symbol	Semantics
$\gamma(Attribute, D)$	Dependency of attribute "D" on attribute "*Attribute*"
"Attribute"	Name of current attribute under consideration
D	Decision attribute (decision class)
γ'_k	Dependency value contributed by object "k" for attribute "*Attribute*"
$V_{\text{Attribute},k}$	Value of attribute "*Attribute*" for object "k" in dataset
n	Total number of previous occurrences of $V_{\text{Attribute},k}$
N	Total number of records in dataset

"positive region" class, so we will add "1" for both. Reading "D", the value "L" now leads to decision class "2" (previously its one occurrence was leading to decision class "1"), so it belongs to "boundary region" class, and thus we will add the value "-1" to overall dependency. Reading object "E" at this stage, value "M" belongs to the "boundary region" class and it had two occurrences before, so, we will add "-2" in the overall dependency. Reading object "F", note that it has already led to "boundary region" class, so we will add "0" to the overall dependency and so on.

$$\gamma(\{a'\}, D) = \frac{1}{10}\sum_{k-1}^{10}\gamma'_k = (1+1+1+(-1)+(-2)+0+1+1+(-2)+0) = \frac{0}{10} = 0$$

Similarly dependency of "D" on {b′, c′, d′} will be as follows:

$$\gamma(\{b'\}, D) = \frac{1}{10}\sum_{k=1}^{10}\gamma'_k = (1+1+1+1+1+(-2)+0+1+(-1)+0) = \frac{3}{10}$$

$$\gamma(\{c'\}, D) = \frac{1}{10}\sum_{k=1}^{10}\gamma'_k = (1+1+1+(-1)+1+0+0+0+1+(-2)) = \frac{2}{10}$$

$$\gamma(\{d'\}, D) = \frac{1}{10}\sum_{k=1}^{10}\gamma'_k = (1+1+1+(-2)+0+1+1+1+(-1)+0) = \frac{3}{10}$$

Note that if a value of an attribute has already led to the boundary region class, then adding a same value will simply be reflected by adding "0" as dependency.

4.3.2 *Direct Dependency Classes (DDC)*

Direct dependency classes are alternate to IDC for calculating dependency directly without involving positive region and exhibit almost the same performance as shown by IDC. DDC determines the number of unique/non-unique classes in a dataset for an attribute C. *A unique class represents the attribute values that lead to*

Table 4.14 How DDC calculates dependency

Dependency	No of unique/non-unique classes
0	If there is no unique class
1	If there is no non-unique class
n	Otherwise where 0 < n < 1

a unique decision class throughout the dataset, so this value can be used to precisely define decision class.

For example, in Table 4.8 the value "L" of attribute b′ is a unique class as all of its occurrences in the same table lead to a single/unique decision class (i.e. "1"). *Non-unique classes represent the attribute values that lead to different decision classes, so they cannot be precisely used to determine the decision class.* For example, in Table 4.8, the value "R" of attribute b′ represents non-unique class as some of its occurrences lead to decision class "2" and some occurrences lead to decision class "3".

Calculating unique/non-unique classes directly lets us avoid complex computations of positive region. The basic idea behind the proposed approach is that the number of unique classes increases dependency and non-unique classes decrease dependency. For a decision class D, the dependency K of D on C is shown in Table 4.14.

The dependency using DDC approach can be calculated by the following formula:

For unique dependency classes:

$$\gamma(Attribute, D) = \frac{1}{N}\sum_{i=1}^{N}(\gamma_i') \tag{4.11}$$

$$\gamma_i' = \begin{cases} 1, & \textit{if class is unique} \\ 0, & \textit{if class is non-unique} \end{cases}$$

And for non-unique dependency classes:

$$\gamma(Attribute, D) = 1 - \frac{1}{N}\sum_{i=1}^{N}(\gamma_i') \tag{4.12}$$

$$\gamma_i' = \begin{cases} 0, & \textit{if class is unique} \\ 1, & \textit{if class is non-unique} \end{cases}$$

Table 4.15 represents the symbol and its semantics.

4.3.2.1 Example

We consider the decision system given in Table 4.7. As per definitions of unique dependency classes,

Table 4.15 Symbols and their semantics

Symbol	Semantics
$\gamma(Attribute, D)$	Dependency of attribute "D" on attribute "*Attribute*"
"Attribute"	Name of current attribute under consideration
D	Decision attribute (decision class)
γ'_i	Dependency value contributed by object "I" for attribute "*Attribute*"
N	Total number of records in dataset

$$\gamma(Attribute, D) = \frac{1}{N}\sum_{i=1}^{N}(\gamma'_i)$$

If attribute {b′} is considered, there are three unique dependency classes, i.e. there are three occurrences of value "L" that lead to unique decision class, so

$$\gamma(\{b'\}, D) = \frac{1}{10}\sum_{i=1}^{10}(\gamma'_i)$$

$$\gamma(\{b'\}, D) = \frac{1}{10}(1 + 1 + 1 + 0 + 0 + 0 + 0 + 0 + 0 + 0) = \frac{3}{10}$$

Similarly for attribute {c′},

$$\gamma(\{c'\}, D) = \frac{1}{10}\sum_{i=1}^{10}(\gamma'_i)$$

$$\gamma(\{c'\}, D) = \frac{1}{10}(0 + 1 + 1 + 0 + 0 + 0 + 0 + 0 + 0 + 0) = \frac{2}{10}$$

On the other hand, if we consider non-unique dependency classes,

$$\gamma(Attribute, D) = 1 - \frac{1}{N}\sum_{i=1}^{N}(\gamma'_i)$$

If we consider attribute "b′", note that there are seven non-unique dependency classes (four occurrences of value "R" lead to two decision classes, and three occurrences of value "N" lead to three decision classes), so

$$\begin{aligned}\gamma(\{b'\}, D) &= 1 - \frac{1}{10}\sum_{i=1}^{10}(\gamma'_i)\\ &= 1 - \frac{1}{10}(0 + 0 + 0 + 1 + 1 + 1 + 1 + 1 + 1 + 1) = \frac{3}{10}\end{aligned}$$

Similarly

$$\gamma(\{c'\},D) = 1 - \frac{1}{10}\sum_{i=1}^{10}(\gamma'_i)$$
$$= 1 - \frac{1}{10}(1+0+0+1+1+1+1+1+1+1) = \frac{2}{10}$$

Note that there are three unique decision classes in attribute {b′} and seven in {c′}. For a decision system,

No of unique classes + no of nonunique classes = size of universe

So we either need to calculate the number of unique classes or non-unique classes. The algorithm for DDC is shown in Fig. 4.6.

Grid is the main data structure used to calculate dependency directly without using positive region. It is a matrix with the following dimensions:

No. of rows = no. of records in dataset
No. of columns = number of conditional attributes + number of decision attributes + 2

So if there are ten records in the dataset, five conditional attributes and one decision class, then Grid dimension will be 10 × 8, i.e. ten rows and eight columns. A row read from the dataset will first be stored in Grid if it does not already exist. The five conditional attributes will be stored in the first five columns; decision attribute will be stored in the sixth column called DECISIONCLASS. The seventh column called INSTANCECOUNT will store the number of times that the record appears in actual dataset; finally the last column called CLASSSTATUS will store the uniqueness of the record; the value "0" mean record is unique and "1" means it is non-unique. If a record, read from the dataset, already exists in Grid, its INSTANCECOUNT will be incremented. If the decision class of the record is different from that already stored in Grid, i.e. the same values of attributes now lead to different decision classes, CLASSSTATUS will be set to "1". However, if the record is inserted for the first time, INSTANCECOUNT is set to "1" and CLASSSTATUS is set to 0, i.e. it is considered unique.

The function "FindNonUniqueDependency" is the main function to calculate the dependency. The function inserts the first record in Grid and then searches for the same record in the entire dataset. The status of the record is updated in Grid as soon as further occurrences of the same record are found. Finally the function calculates dependency value on the basis of uniqueness/non-uniqueness of classes. Function "InsertInGrid" inserts the record in the next row of the Grid. "FindIndex" finds the row no. of the current record in the Grid. "IfAlreadyExistsInGrid" finds if the record already exists or not. Finally "UpdateUniquenessStaus" function updates the status of the record in Grid.

```
Function FindNonUniqueDependency
Begin
InsertInGrid(X1)
For i=2 to TotalUnivesieSize
IfAlreadyExistsInGrid(Xi)
        Index = FindIndexInGrid(Xi)
        If DecisionClassMatched(index,i) =  False
UpdateUniquenessStaus(index)
        End-IF
   Else
InsertInGrid(Xi)
  End-IF

Dep=0

For i=1 to TotalRecordsInGrid
 If Grid(I,CLASSSTATUS) = 1
   Dep= Dep+ Grid(i,INSTANCECOUNT)
 End-IF
Return (1-Dep)/TotalRecords
End Function

Function InsertInGrid(Xi)
For j=1 to TotalAttributesInX
 Grid(NextRow,j) = Xi^j
End-For
 Grid(NextRow,DECISIONCLASS) = Di
Grid(NextRow, INSTANCECOUNT) = 1
Grid(NextRow, CLASSSTATUS) = 1 // 1 => uniqueness
End-Function

Function IfAlreadyExistsInGrid(Xi)
 For i=1 to TotalRecordsInGrid
    For j=1 to TotalAttributesInX
     If Grid(i,j) <>X^j
          Return False
    End-For
 End-For
Return True
End-Function

Function FindIndexInGrid(Xi)
 For i=1 to TotalRecordsInGrid
RecordMatched=TRUE
    For j=1 to TotalAttributesInX
     If Grid(i,j) <>X^j
RecordMatched=FALSE
    End-For
  If RecordMatched= TRUE
    Return j
  End-If
 End-For
Return True
End Function

Function DecisionClassMatched(index,i)
  If Grid(index, DECISIONCLASS) = Di
    Return TRUE
  Else
    Return False
  End-If
End-Function

Function UpdateUniquenessStaus(index)
Grid(index, CLASSSTATUS) = 1
End-Function
```

DDC using non-unique classes

Fig. 4.6 Pseudocode for directly finding dependency using unique and non-unique classes

4.4 Redefined Approximations

Unique decision classes lead to the idea of calculating lower and upper approximation without using indiscernibility relation [3]. Calculating lower and upper approximation requires calculating equivalence classes (indiscernibility relation) which is computationally an expensive task. Using unique decision classes lets us avoid this task and we can directly calculate the lower approximation. The new definitions are semantically same to the conventional definitions but provide computationally more efficient method for calculating these approximations by avoiding equivalence class structure. The following section discusses the proposed new definitions in detail.

4.4.1 Redefined Lower Approximation

The conventional rough set-based lower approximation defines the set of objects that can with certainty be classified as members of concept "X". For attribute (s) $c \in C$ and concept X, the lower approximation will be

$$\underline{C}X = \{X | [X]_c \subseteq X\} \tag{4.13}$$

This definition requires calculation of indiscernibility relation, i.e. equivalence class structure $[X]_c$, where the objects belonging to one equivalence class are indiscernible with respect to the information present in attribute(s) $c \in C$. Based on the concept of lower approximation provided by RST, we have proposed a new definition as follows:

$$\underline{C}X = \{\forall x \in U, c \in C, a \neq b | \, x_{\{c \cup d\}} \rightarrow a, x_{\{c \cup d\}} \nrightarrow b\} \tag{4.14}$$

i.e., the lower approximation of concept "X" w.r.t. the attribute set "c" is a set of objects such that for each occurrence of the object, the same value of conditional attribute "c" always leads to the same decision class value. So, if there are "n" occurrences of an object, then all of them lead to the same decision class (for the same value of attributes), which alternatively means that for a specific value of an attribute, we can with surety say that the object belongs to a certain decision class. This is exactly equal to conventional definition of lower approximation.

So, the proposed definition is semantically the same as the conventional one; however, computationally it is more convenient for calculating lower approximation; it avoids construction of equivalence class structures to find the objects belonging to positive region. It directly scans the dataset and finds the objects that lead to the same decision class throughout, thus enhancing the performance of algorithm using this measure. We will use Table 4.16 as sample to calculate lower approximation using both definitions.

Table 4.16 Sample decision system

U	a	b	c	d	Z
X1	L	3	M	H	1
X2	M	1	H	M	1
X3	M	1	M	M	1
X4	H	3	M	M	2
X5	M	2	M	H	2
X6	L	2	H	L	2
X7	L	3	L	H	3
X8	L	3	L	L	3
X9	M	3	L	M	3
X10	L	2	H	H	2

We suppose the concept $X = \{x|Z(x) = 2\}$, i.e. we will find the objects about which could with surety say that they lead to decision class "2". Conventional definition requires three steps given below:

Step 1 Calculate the objects belonging to the concept X. Here is our case concept $X = \{x \mid Z(x) = 2\}$.

So, we will identify the objects belonging to the concept. In our case

$$\mathrm{X} = \{X4, X5, X6, X10\}$$

Step 2 Calculate equivalence classes using conditional attributes.

Here we will consider only one attribute "b" for simplicity, i.e. $C = \{b\}$. So, we will calculate equivalence classes using attribute "b", which in our case becomes

$$[X]_c = \{X1, X4, X7, X8, X9\}\{X2, X3\}\{X5, X6, X10\}$$

Step 3 Find objects belonging to lower approximation.

In this step we actually find the objects that belong to lower approximation of the concept w.r.t. to considered attribute. Mathematically, this step involves finding objects from $[X]_c$ which are subset of X, i.e. $\{[X]_c \subseteq X\}$. In our case,

$$\underline{C}X = \{[X]_c \subseteq X\} = \{\mathrm{X5, X6, X10}\}$$

Using the proposed definition, we construct the lower approximation without using equivalence class structures. We directly find the objects that under the given concept always lead to same decision class (concept) for the current value of attributes.

In our case, we just pick an object and find if for the same values of attributes, it leads to some other decision class or not. We find that objects X5, X6 and X10 always lead to the same decision class, i.e. the concept under consideration for attribute "b". On the other hand, objects X2 and X3 do not lead to concept X, so they will not be part of lower approximation. Similarly some occurrences of objects X1, X4, X7, X8 and X9 also lead to a different decision class than X, so they are also excluded from lower approximation. So, we can with surety say that objects X5, X6 and X10 belong to the lower approximation.

As discussed earlier, semantically both definitions are the same, but computationally the proposed definition is more effective as it avoids construction of equivalence class structure. It directly calculates the objects belonging to lower approximation by checking objects that always lead to same decision class under consideration. Directly calculating the lower approximation in this way lets us exclude complex equivalence class structure calculation which makes algorithms using this measure more efficient. Using conventional definition on the other hand requires three steps discussed in the previous section. Performing these steps offers a significant performance bottleneck to algorithms using this measure.

4.4.2 Redefined Upper Approximation

Upper approximation defines the set of objects that can only be classified as possible members of X w.r.t. the information contained in attribute(s) $c \in C$. For attribute(s) $c \in C$ and concept X, the lower approximation will be

$$\bar{C}X = \{X | [X]_B \cap X \neq 0\}$$

i.e., upper approximation is the set of objects that may belong to the concept X w.r.t. information contained in attribute(s) c.

On the bases of the same concept, a new definition of upper approximation was proposed as follows:

$$\bar{C}X = \underline{C}X \cup \{\forall x \in U c \in C a \neq b\, x_{\{c\}} \rightarrow a x'_{\{c\}} \rightarrow b\}$$

This definition will be read as follows:

Provided that objects x and x' are indiscernible w.r.t. to attribute(s) c, they will be part of an upper approximation if either they belong to lower approximation or at least one of their occurrences leads to decision class belonging to concept X. So objects x and x' belong to upper approximation if both occurrences of them lead to a different decision class for the same value of attributes. For example, in Table 4.16 the objects X1, X4, X7, X8 and X9 lead to a different decision class for the same value of attribute "b". So all of them can be possible members of upper approximation of concept Z = 2.

As with redefined lower approximation, the proposed definition for the upper approximation is semantically the same as conventional upper approximation. However, it helps us directly calculate objects belonging to upper approximation without calculating the underlying equivalence class structures.

Like the conventional definition of lower approximation, calculating upper approximation requires three steps again. We will again use Table 4.16 as sample to calculate upper approximation using both definitions. We suppose the concept: $X = \{x \mid Z(x) = 2\}$ as shown in example of lower approximation.

Step 1 Calculate the objects belonging to the concept X. Here is our case concept $X = \{x \mid Z(x) = 2\}$.

We have already performed this step and calculated the objects belonging to the concept X on the basis of a decision class in the previous example. So

$$X = \{X4, X5, X6, X10\}$$

Step 2 Calculate equivalence classes using conditional attributes.

This step has also been calculated before, and objects belonging to $[X]_c$ were identified as follows by considering attribute "b":

$$[X]_c = \{X1, X4, X7, X8, X9\}\{X2, X3\}\{X5, X6, X10\}$$

Step 3 Find the objects belonging to upper approximation.

In this step we finally calculate the objects belonging to lower approximation. The step is calculated by identifying the objects in $[X]_c$ that have non-empty interaction with objects belonging to concept X. The intention is to identify all those objects at least one instance of which leads to the decision class belonging to the concept X. Mathematically $[X]_c \cap X \neq 0$. In our case,

$$\bar{C}X = \{X1, X4, X7, X8, X9\}\{X5, X6, X10\}$$

Proposed definition of upper approximation avoids calculating computational expensive step of indiscernibility relation. It simply scans the entire dataset for the concept X using the attributes c and finds all the indiscernible objects such that any of their occurrence belongs to the concept X. At least one occurrence should lead to a decision class belonging to concept X. In our case, objects X1, X4, X7, X8 and X9 are indiscernible and at least one of their occurrences leads to concept Z=2. Objects X2 and X3 are indiscernible, but none of their occurrence leads to required concept, so they will not be part of upper approximation. Objects X5, X6 and X10 belong to lower approximation so they will be part of upper approximation as well. In this way using the proposed definition, we find the following objects as part of upper approximation:

$$\bar{C}X = \{X1, X4, X7, X8, X9\}\{X5, X6, X10\}$$

That is the same as produced by using the conventional method.

Semantically proposed definition of upper approximation is the same as conventional one, i.e. it produces the same objects that may possibly be classified as members of concept X. However, it calculates these objects without calculating equivalence classes. So, the proposed approach is computationally less expensive which consequently can result in enhanced performance of the algorithms using proposed method. The conventional definition on the other hand again requires three steps (as discussed above) to calculate upper approximation which impose computational implications.

When a problem in real world becomes so complex that you cannot decide a problem to be true or false or you cannot predict the accuracy of a statement to be one or zero, in that case we say its probability is uncertain. Normally we solve the problems through probability, but when you get a system where all elements are vague and not clear, then we use fuzzy theory to handle such systems. Fuzzy means vague, something that is not clear.

Now what is the difference between fuzzy and crisp set? We can explain with simple argument. If something can be answered with yes or no, then it means it is crisp, e.g. is water colorless? Definitely the answer will be yes, so this statement is crisp, but what about the statement is John honest? Now you cannot precisely answer this question because someone would think that John is more honest, yet some other would take him less honest, etc., so, this statement will not be crisp but fuzzy.

Now the question is that why does fuzzy situation arise? It arises due to our partial information about the problem or due to information that is not fully reliable or inherent imprecision in statement due to which we will say that problem is fuzzy.

4.5 Summary

In this chapter we have presented some advance concepts of rough set theory, and we discussed advance forms of three fuzzy concepts, i.e. lower approximation, upper approximation and dependency. Then some preliminary concepts regarding fuzzy set theory, hybridization of RST with fuzzy set theory, were presented.

Bibliography

1. Raza MS, Qamar U (2016) An incremental dependency calculation technique for feature selection using rough sets. Inform Sci 343:41–65
2. Al Daoud E (2015) An efficient algorithm for finding a fuzzy rough set reduct using an improved harmony search. Int J Modern Educ Comput Sci 7.2:16

3. Raza MS and U Qamar (2016) Feature selection using rough set based dependency calculation. Phd Dissertation, National University of Science and Technology
4. Lingras P and R Jensen (2007) Survey of rough and fuzzy hybridization. Fuzzy Systems Conference. FUZZ-IEEE 2007. IEEE International. IEEE, 2007
5. Sarkar M (2000) Fuzzy-rough nearest neighbors algorithm. Systems, Man, and Cybernetics, IEEE International Conference on. Vol. 5. IEEE, 2000
6. De Cock M and C Cornelis (2005) Fuzzy rough set based web query expansion. In: Proceedings of Rough Sets and Soft Computing in Intelligent Agent and Web Technology, International Workshop at WIIAT2005
7. Jensen R, Shen Q (2005) Fuzzy-rough data reduction with ant colony optimization. Fuzzy Set Syst 149(1):5–20
8. Hu Q, Yu D, Xie Z (2006) Information-preserving hybrid data reduction based on fuzzy-rough techniques. Pattern Recogn Lett 27(5):414–423
9. Dubois D, Prade H (1990) Rough fuzzy sets and fuzzy rough sets. Int J Gen Syst 17 (2-3):191–209
10. Jensen R and Q Shen (2008) Computational intelligence and feature selection: rough and fuzzy approaches, vol 8. Wiley, Hoboken
11. Hu Q, Zhang L, Zhou Y, Pedrycz W (2017) Large-scale multi-modality attribute reduction with multi-Kernel fuzzy rough sets. IEEE Transactions on Fuzzy Systems
12. Jensen R, Mac Parthaláin N (2015) Towards scalable fuzzy–rough feature selection. Inform Sci 323:1–15
13. Qian Y, Wang Q, Cheng H, Liang J, Dang C (2015) Fuzzy-rough feature selection accelerator. Fuzzy Sets Syst 258:61–78
14. Jensen R, Vluymans S, Mac Parthaláin N, Cornelis C, Saeys Y (2015) Semi-supervised fuzzy-rough feature selection. In: Rough sets, fuzzy sets, data mining, and granular computing. Springer International Publishing, pp 185–195
15. Kuzhali JV, Rajendran G, Srinivasan V, Kumar GS (2010) Feature selection algorithm using fuzzy rough sets for predicting cervical cancer risks. Modern Appl Sci 4(8):134

Chapter 5
Rough Set-Based Feature Selection Techniques

Rough set theory has been successfully used for feature selection techniques. The underlying concepts provided by RST help find representative features by eliminating the redundant ones. In this chapter, we will present various feature selection techniques which use RST concepts.

5.1 QuickReduct

In QuickReduct (QR) [1], authors have proposed a feature selection algorithm using rough set theory. Algorithm uses forward feature selection to find reducts without exhaustively generating all possible feature subsets. Starting with an empty set, algorithm keeps on adding attributes one by one which results in a maximum increase in the degree of dependency. Algorithm continues until maximum dependency value is achieved. After adding each attribute, dependency is calculated and attribute is kept if it results in maximum increase in dependency. If the value of current feature subset becomes the same as that of entire set of conditional attributes, algorithm terminates and outputs current set as reduct. Figure 5.1 shows the pseudocode of the algorithm.

Algorithm uses positive region-based approach for calculating dependency at Steps 5 and 8. However, using positive region-based dependency measure requires three steps, i.e. calculating equivalence classes using decision attribute, calculating dependency classes using conditional attributes and finally calculating positive region. Using these three steps can be computationally expensive.

As QR is one of the most common RST-based feature selection algorithms, we explain it with the help of an example, but first we see how does it work? Algorithm starts with an empty reduct set R; it then uses a loop to calculate dependency of each attribute subset. Initially subset comprises a single attribute. The attribute subset with maximum dependency is found at the end of each iteration and is considered as

M.S. Raza, U. Qamar, *Understanding and Using Rough Set Based Feature Selection: Concepts, Techniques and Applications*,
DOI 10.1007/978-981-10-4965-1_5

Fig. 5.1 QuickReduct algorithm

```
QUICKREDUCT(C, D)
C, The set of all conditional features
D, The set of decision features
    R ← {}
    do
    T ← R
    ∀ x ∈ (C − R)
    If γ_{RU{x}}(D) > γ_T(D)
    T ← RU{x}
    R ← T
    until γ_R(D) == γ_c(D)
    Output R
```

reduct set R. For the next iteration, previous attribute subset with maximum dependency is joined with every single attribute in conditional attribute set, and a subset with maximum dependency is found again. The process continues until at any stage an attribute subset with dependency equal to that of the entire conditional attribute set is found. The sequence is as follows:

1. Find dependency of each attribute.
2. Find attribute with maximum dependency and make it R.
3. Start next iteration.
4. Join R with each single attribute in {C}-{R}.
5. Again calculate subset with maximum dependency.
6. Now R comprises two attributes.
7. Again join R with each single attribute in {C}-{R}.
8. Again calculate subset with maximum dependency.
9. Now R comprises three attributes.
10. Repeat the entire process until at any stage dependency of R = 1 or that of the entire set of conditional attributes.

We now explain this with the help of an example. We will consider the following dataset (Table 5.1):

Here C = {a, b, c, d} are conditional attributes whereas D = {Z} is decision class.

Initially R = {ϕ}. Join R with each attribute in conditional set and find the attribute subset with maximum dependency:

$$\gamma(\{R \cup A\}, Z) = 0.1$$
$$\gamma(\{R \cup B\}, Z) = 0.5$$
$$\gamma(\{R \cup C\}, Z) = 0.3$$
$$\gamma(\{R \cup D\}, Z) = 0.0$$

So far {R ∪ B} results in maximum increase in dependency, so R = {b}. Now we join R with each conditional attribute except {b}:

Table 5.1 Sample decision system

U	a	b	c	d	Z
X1	L	3	M	H	1
X2	M	1	H	M	1
X3	M	1	M	M	1
X4	H	3	M	M	2
X5	M	2	M	H	2
X6	L	2	H	L	2
X7	L	3	L	H	3
X8	L	3	L	L	3
X9	M	3	L	M	3
X10	L	2	H	H	2

$$\gamma(\{R \cup A\}, Z) = 0.6$$
$$\gamma(\{R \cup C\}, Z) = 0.8$$
$$\gamma(\{R \cup D\}, Z) = 0.6$$

So, {R ∪ C} results in maximum increase in dependency, so R = {b,c}. Now, again we join R with every conditional attribute except {b, c}:

$$\gamma(\{R \cup A\}, Z) = 1.0$$
$$\gamma(\{R \cup D\}, Z) = 1.0$$

Note that both {R ∪ A} and {R ∪ D} result in the same degree of increase in dependency, so algorithm will pick the first one, i.e. {R ∪ A}, and will output R = {a,b,c} as reduct. Here we find a very interesting aspect of the algorithm that performance of this algorithm depends on distribution of the conditional attributes as well. There are many variations proposed on QR in intention to optimize feature selection process. One is ReverseReduct [1] and the other is accelerated QR [2].

ReverseReduct [1] is the strategy for attribute reduction; however, it uses backward elimination in contrast with forward feature selection mechanism. Initially the entire set of conditional attributes is considered as reduct. It then keeps on removing attributes one by one until no further attribute can be removed without introducing inconsistency. Algorithm also suffers the same problem as that faced by QuickReduct. It uses positive region-based dependency measure and is equally unsuitable for larger datasets.

In accelerated QR, on the other hand, attributes are selected in order of degree of dependency. Algorithm starts by calculating dependency of each conditional attribute and selects the attribute with maximum dependency. In the case of two or more features having the same dependency, they are combined to calculate dependency again. Both the values are then compared. If the condition becomes "false", the attributes are chosen in combination with the next highest degree of dependency

Accelerated Quickreduct(C, D)
C: $c_1, c_2, \ldots, c_n$, the set of all conditional features;
D: d, a decision features

(a) R ← {}
(b) $\gamma_{prev} = 0,\ \gamma_{best} = 0$
(c) Do
(d) T ← R
(e) $\gamma_{prev} = \gamma_{best}$
(f) *Compute* γ_i, i = 1…n where γ_R(D) = card($POS_R(D)$)/card(U)
$$POS_R(D) = \_RX$$
(g) Select Max(γ_i) *or* (γ_j) where $j \in$ { the set of all attributes which have the same highest degree of dependency}
(h) if γ_{RUX}(D) > γ_{prev}(D)
(i) T ← RU{X}
(j) $\gamma_{best} = \gamma_T(D)$
(k) R ← T
(l) until $\gamma_{best} == \gamma_{prev}$
(m) return R

Fig. 5.2 Accelerated QR

value, and the process is repeated until the condition becomes "true". Figure 5.2 shows the pseudocode of accelerated QR algorithm.

5.2 Hybrid Feature Selection Algorithm Based on Particle Swarm Optimization (PSO)

In [3] Hanna et al. proposed a new feature selection algorithm by using hybridization of rough set theory, quick reduct and particle swarm optimization. Hybridization in this way results in benefits of all of these three techniques. Starting with empty set, it keeps on adding attributes one by one. It constructs a population of particles with random position and velocity in S dimensions in the problem space. Fitness of each particle is then computed using RST-based dependency measure. The feature with highest dependency is selected, and the combination of all other features with this one is constructed. Fitness of each of these combinations is selected. If the fitness value of this particle is better than previous best (pbest) value, this is selected as pbest. Its position and fitness are stored. It then compares the fitness of current particle with population's overall previous best fitness (gbest). If particle's fitness is better than previous gbest, it becomes current gbest. Algorithm then updates velocity and position of each particle. It continues until stopping criteria is met which is maximum number of iterations in normal case. According to the algorithm, the dependency of each attribute subset is calculated based on the dependency on decision attribute and the best particle is chosen. Algorithm uses positive region-based dependency measure and is an enhancement of QuickReduct algorithm.

Velocity of each particle is represented using positive number from 1 to V_{max}. It implies how many bits of a particle should be changed to be the same as that of globalbest position. The number of bits different between two particles implies the difference between their positions, e.g. if Pgbest = [1,0,1,1,1,0,1,0,0,1], Xi = [0,1,0,0,1,1,0,1,0,1], then the difference between Pgbest and Xi is Pgbest − Xi = [1,−1,1,1,0,−1,1,−1,0,0]. "1" means that this bit (each "1" represents the presence of feature and "0" represents absence) should be selected as compared to the globalbest position, and "-1" means that this bit should not be selected. After velocity is updated, the next task is to update position by new velocity. If the new velocity is V and the number of different bits between the current particle and gbest is xg, then position is updated as per the following conditions:

- V ≤ xg, then random V bits, different than that of gbest, are changed. This will make particle move towards the best position.
- V > xg. In this case, (V − xg) further bits should also be randomly changed. This will make the particle move some distance towards other directions for further exploration ability.

Figure 5.3 shows the pseudocode of PSO-QR algorithm.

5.3 Genetic Algorithm

In [4] authors have a genetic algorithm using rough set theory. Algorithm uses positive region-based dependency measure to calculate fitness of each chromosome where each chromosome comprises genes representing the presence of attributes from dataset. The selected feature subset is provided to classify for further analysis. Chromosomes with the highest fitness degree are considered to be candidates for selection. Stopping criterion was defined based on the following equation:

$$k = \gamma(C,D) = \frac{|POS_C(D)|}{|U|} \geq \alpha$$

The candidates equal or greater than were accepted as result. The following equation was used to calculate solution addition type added to the solution:

$$RSC\% = 100\% - (BSC\% + WSC\%)$$

where:

RSC = random selected chromosomes
BSC = best solution candidates
WSC = worst solution candidates

```
Input: C, the set of all conditional features,
       D, the set of decision features.
Output: Reduct R

Step 1: Initialize X with random position Vi with random velocity
∀: Xi ← randomPosition();
Vi ← randomVelocity();
Fit ← 0; globalbest ← Fit;
Gbest ← X1; Pbest(1) ← X1
For i = 1 … S
pbest(i) = Xi
Fitness(i) = 0
End For

Step 2: While Fit != 1   // stopping criterion
For i = 1 … S  // for each particle
∀: Xi;
//Compute fitness of feature subset of Xi
R ← Feature subset of Xi (1's of Xi)
∀ x ∈ (C − R)
γRU(X)(D) = |POSRU(X)(D)| / |U|

Fit = γRU(X)(D) ∀ xcR, γx(D) ≠ γc(D)

End For

Step 3: Compute best fitness
For i = 1:S
if(Fitness(i) > globalbest)  // if current fitness is greater than global best fitness
globalbest ← Fitness(i); // assign current fitness value as global best fitness
gbest ← Xi ;
getReduct(Xi)
Exit
End if
End for
UpdateVelocity(); // Update velocity Vi's of Xi's
UpdatePosition(); // Update position of Xi's
// Continue with the next iteration
End {while}
Output Reduct R
```

Fig. 5.3 PSO-QR

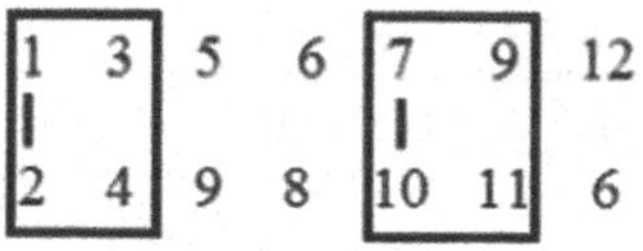

Fig. 5.4 Selected chromosomes for order-based crossover method

A number of generations in each generation pool were 2*n, where "n" is user-defined parameter and can be changed for optimizing the performance. It specifies number of generations. In proposed version, these 2*n (2, 4, 6... n) generations were randomly initialized and used for generating the following generations. The last 2*n (4, 6, 8...) generations were used to construct the gene pool used to determine the intermediate region used for crossover and mutation operator.

Order-based and partially matched crossover methods were used to perform crossover operator. In order-based method, a random number of solution points are selected from parent chromosomes. In the first chromosome, the selected gene will remain at its place, whereas in the second chromosome, the corresponding gene will be beside that of the first chromosome that occupies the same place. Order-based crossover method is shown in Figs. 5.4 and 5.5. Figure 5.4 shows the selected chromosomes and Fig. 5.5 shows the resultant chromosomes.

In partially matched method (PMX), two crossover points are randomly selected to give matching selection. Position-wise exchange takes place then. It affects cross by position-by-position exchange operations. It is also called partially mapped crossover as parents are mapped to each other. Figures. 5.6 and 5.7 show the process of partially mapped crossover method.

For mutation, inversion and two-change mutation operators were used. In inversion method, we randomly select a subtour by determining two points in chromosome and invert genes between selected points, whereas in adjacent two-input change mutation method, two adjacent genes are selected and their places are inverted. Figures 5.8 and 5.9 show both mutation methods.

5.4 Incremental Feature Selection Algorithm (IFSA)

Qian et al. [5] presented a new feature selection algorithm for dynamic datasets where records and attributes keep on adding from time to time. Initially P is assigned with original feature subset, and then new attributes are added; dependency of P is computed again, if it is equal to that of the entire attribute set, P that is the new feature subset is computed. Algorithm uses the following definitions to measure significance of an attribute:

Definition 1 Let DS $= (U, A = C \cup D)$ be a decision system, for $B \subseteq C$ and $a \in B$. The significance measure of attribute "a" is defined by $sig_1(a,B,D) = \gamma B(D) - \gamma_{B-\{a\}}(D)$.

If $sig_1(a,B,D) = 0$, then the feature "a" can be removed, otherwise not.

1 2 5 6 7 10 12

2 1 9 8 10 7 6

Fig. 5.5 Selected chromosomes for order-based crossover method

1 2 3 | 5 4 6 7 | 8 9

4 5 2 | 1 8 7 6 | 9 3

Fig. 5.6 Selected chromosomes for partially mapped crossover method

8 1 2 | 5 4 6 7 | 9 3

5 2 3 | 1 8 7 6 | 4 9

Fig. 5.7 Chromosomes resulted after crossover operator

2 5 6 | 8 10 12 14 | 15 17

2 5 6 | 14 12 10 8 | 15 17

Fig. 5.8 Inversion mutation method

2 5 6 [8 10] 12 14 15 17

2 5 6 [10 8] 12 14 15 17

Fig. 5.9 Adjacent two-change mutation method

Definition 2 Let DS = (U,A = C ∪ D) be a decision system, for B ⊆ C and a/∈ B. The significance of feature "a" is defined by sig_2(a,B,D) = γB∪{a}(D) − γB(D).

Figure 5.10 shows the pseudocode of the algorithm.

5.5 Feature Selection Method Using Fish Swarm Algorithm (FSA)

Chen et al. [6] presented a new fish swarm algorithm based on rough set theory. Algorithm constructs initial swarm of fish where each fish represents a feature subset. With passage of time, these fish search for food, change their position for this purpose and communicate with each other to find a locally and globally best position, the position with minimum high density of food. Once a fish having maximum fitness is found, algorithm perishes it by outputting the reduct it represents. When all the fish are perished, next iteration starts. Process continues until it gets the same reducts in three consecutive iterations or maximum iteration threshold is met. Figure 5.11 shows the flow of FSA process.

Some underlying concepts that must be considered before applying FSA to feature selection are:

Input A decision system $DS = (U, A = C \cup D)$, the original feature subset Red, the original position region $\gamma c(D)$, and the adding feature set C_{ad} or the deleting feature set C_{de}, where $C_{ad} \cap C = \emptyset$, $C_{ad} \subseteq C$;
Output A new feature subset Red'.

Begin
1) Initialize $P \leftarrow Red$;
2) If a feature set C_{ad} is added into the system DS;
3) Let $C' \leftarrow C \cup C_{ad}$;
4) Compute the equivalence classes U/C' and $\gamma c'(D)$;
   //According to Theorem 1
5) for $i = 1$ to $|C_{ad}|$ do
6) compute $sig1(c_i, C_{ad}, D)$;
7) if $sig1(c_i, C_{ad}, D) > 0$, then $P \leftarrow P \cup \{c_i\}$;
8) end for
9) if $\gamma p(D) = \gamma c'(D)$, turn to Step 25; else turn to Step 16;
10) End if
11) If a feature set C_{de} are deleted from the system DS;
12) Let $C' \leftarrow C - C_{de}$;
13) if $C_{de} \cap P = \emptyset$, turn to step 25; else $P \leftarrow P - C_{de}$ and turn to step 14;
14) Compute the equivalence classes U/C' and $\gamma c'(D)$;
   //According to Theorem 2
15) End if
16) For $\forall c \in C' - P$, construct a descending sequence by $sig2(c, P, D)$, and record the result by $\{c'_1, c'_2, \ldots, c'_{|C'-P|}\}$;
17) While $(\gamma p(D) \neq \gamma c'(D))$ do
18) for $j = 1$ to $|C' - P|$ do
19) select $P \leftarrow P \cup \{c'_j\}$ and compute $\gamma p(D)$;
20) End while
21) For each $c_j \in P$ do
22) compute $sig1(c_j, P, D)$;
23) if $sig1(c_j, P, D) = 0$, then $P \leftarrow P - \{c_j\}$;
24) end for
25) $Red' \leftarrow P$, return Red';

End

Fig. 5.10 IFSA

5.5.1 Representation of Position

A fish position is represented by binary bit string of length N, where N is the total number of features. The presence of a feature in a fish is represented by binary bit "1" and absence of a feature is represented by "0". For example, if $N = 5$, the following fish shown in Fig. 5.12 represents the presence of the first, third and fourth feature from dataset.

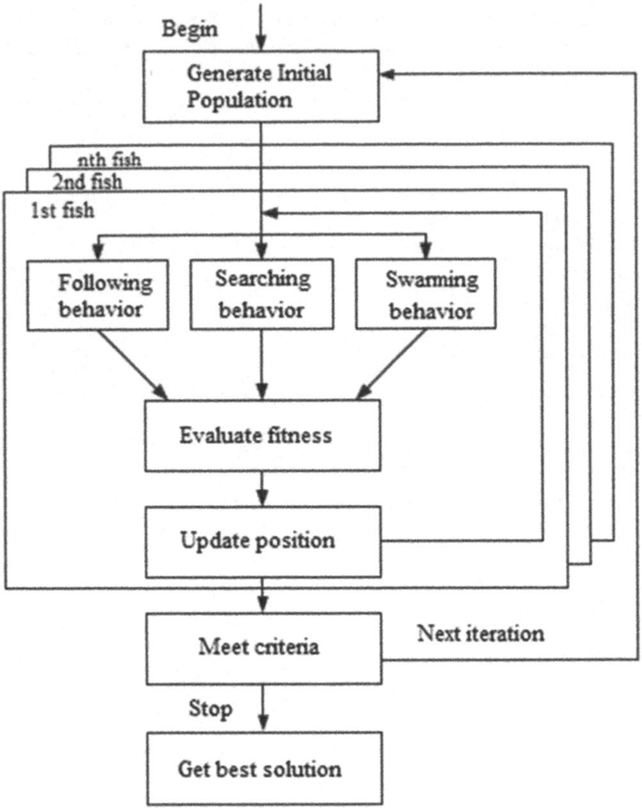

Fig. 5.11 Flow of FSA

Fig. 5.12 Sample fish

1	0	1	1	0

5.5.2 *Distance and Centre of Fish*

Suppose two fish are represented by two bit strings X, Y representing the position of these two fish; hamming distance will be calculated by X XOR Y, i.e. the number of bits at which strings are different. Mathematically,

$$h(X, Y) = \sum_{i=1}^{N} x_i \bigoplus y_i,$$

where "$\bigoplus$" is modulo-2 addition, x_i ; y_i {0, 1}. The variable x_i represents a binary bit in X.

5.5.3 Position Update Strategies

In each iteration, every fish starts with a random position. Fish change their position one step according to searching, swarming and following behaviour. Authors have used fitness function to evaluate all of these behaviours. The behaviour with maximum fitness value updates the next position.

5.5.4 Fitness Function

The following fitness function was used in the algorithm:

$$\text{Fitness} = \alpha^{*}\gamma_R(D) + \beta^{*}\frac{|C| - |R|}{|C|},$$

where $\gamma_R(D)$ is the dependency of decision attribute "D" on "R", R is the number of "1" bits in a fish and |C| is the number of features in dataset.

5.5.5 Halting Condition

Whenever a single fish finds maximum fitness, it is perished by getting reduct. Algorithm terminates when the same reduct is obtained in three consecutive iterations or maximum number of iterations are met.

5.6 Feature Selection Method Based on QuickReduct and Improved Harmony Search Algorithm (RS-IHS-QR)

Inbarani et al. [7] propose an improved harmony search-based feature selection algorithm using rough set theory. Algorithm operates in five steps. In the first step, it initializes all the variables and parameters. Algorithm uses lower approximation-based dependency to measure fitness of a memory vector. It then initializes harmony memory. Memory is initialized randomly by the attributes present in dataset. It is a vector having the size of HMS where each row in this vector represents a possible solution. In the third step, it improvises new harmony vector based on a harmony memory consideration rate, pitch rate, adjustment rate and random function. In the fourth step, it updates worst harmony memory vector with the one having better fitness. Finally algorithm stops, in the fifth step, if stopping criteria are satisfied.

5.7 A Hybrid Feature Selection Approach Based on Heuristic and Exhaustive Algorithms Using Rough set Theory (FSHEA)

In [8] we have proposed a new solution for feature selection based on relative dependency. Their approach is a two-step process:

Step 1: A preprocessor that finds initial feature subset. For this step, we have used genetic algorithm and PSO for the sake of experiments. Any heuristic algorithm, e.g. FSA, ACO, etc., can be used.

Step 2: Feature subset optimization. This step optimizes the initial solution produced by Step 1, by removing unnecessary features. Here, any of the exhaustive feature subset algorithms can be used. We have used the one based on relative dependency, as it avoids the calculation of computationally expensive positive region.

Figure 5.13 shows the diagrammatic representation of the proposed solution.

The proposed approach has changed the role of the algorithms used in any of the above-mentioned steps. Instead of being used to find the actual reduct set, heuristic algorithms find an initial reduct set, which is then refined by using exhaustive search. The following section provides brief description of each of these methods along with their role in hybrid approach.

5.7.1 Feature Selection Preprocessor

Preprocessor is the first step of the proposed solution which provides us with initial candidate reduct set. At this stage, we have used heuristic algorithms. The reason behind is that heuristic algorithms help us find initial feature subset within minimal time without exhaustively searching the entire dataset. However, heuristic algorithms do not ensure optimal solution, so the feature subset produced by preprocessor may contain unnecessary features as well, which are then removed in optimization step. At this stage, any type of heuristic algorithms can be used. For the sake of this paper, we have used genetic algorithm and particle swarm optimization approach. Below is the brief description of both of these algorithms along with slight changes made for the purpose of enhancing computational efficiency.

5.7.1.1 Genetic Algorithm (GA) as Preprocessor

Genetic algorithm is one of the heuristic-based algorithms where candidate solutions are deployed in the form of chromosomes. Each of the chromosomes comprises the genes. In case of feature subset selection, a gene may represent the

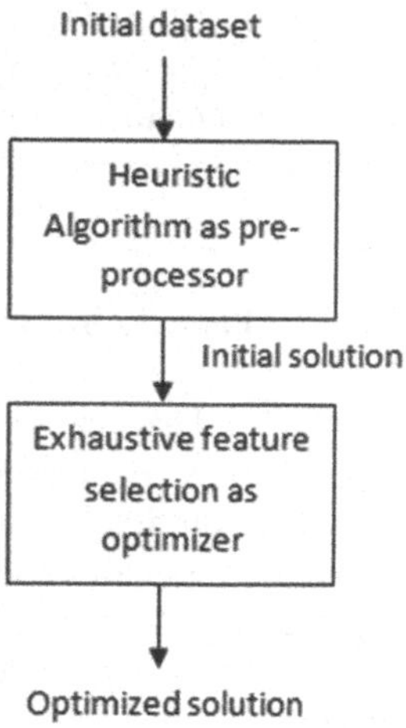

Fig. 5.13 Proposed hybrid approach for feature selection

presence of an attribute. Complete detail of GA can be found in [8]. Here is description of the points where we made slight changes in this algorithm.

Chromosome Encoding In conventional genetic algorithm, the chromosome comprises the set of genes, each of which is randomly selected and represents a part of the solution. We, however, have avoided the random gene selection. Instead chromosomes were encoded in such a way that all the attributes of the datasets are present, e.g. if dataset has 25 attributes, then a particular generation may have 5 chromosomes each having 5 genes and each gene representing an attribute. This step was taken for two reasons:

(a) Selecting the chromosome size is a problem in GA. To let GA be independent of the size, we encoded all the attributes as genes so that if any size of chromosome is chosen, all attributes are tested.
(b) As genes are randomly selected, so there is a chance that some attributes are skipped. Encoding all the attributes ensures that all the attributes are tested.

Note that any other encoding scheme can be used in the proposed hybrid approach.

Crossover For the purpose of our algorithm, we have used conventional one-point crossover. However, crossover was performed in decreasing order of relative dependency. This was based on our observation that if a chromosome C_1 with higher order of relative dependency crosses over with another higher-order relative dependency chromosome, it is more likely that the produced offspring will have high relative dependency value as compared to the one if C_1 crosses over with a lower relative dependency value chromosome. Using the above-mentioned crossover order is more likely to produce higher relative dependency offspring, thus resulting in fewer number of generations.

Fitness Function The fitness function comprised the relative dependency of the attributes represented by genes present in current chromosome. Relative dependency was calculated by using equation 7. The first chromosome with the highest fitness value, i.e. one (1), was chosen as candidate solution.

5.7.1.2 Particle Swarm Optimization (PSO) Algorithm as Preprocessor

We also used particle swarm algorithm as preprocessor. It is another heuristic-based approach based on swarm logic. A particle in a swarm represents a potential candidate solution. Each swarm has a localbest represented by pbest.

Algorithm evaluates fitness of all particles and the one having the best fitness becomes localbest. After finding localbest, PSO attempts to find globalbest particle, i.e. gbest. Gbest particle is the one having the best fitness so far throughout the swarm. PSO then attempts to update position and velocity of each particle based on its position and distance from localbest. Algorithm can be deployed for feature selection using rough set-based conventional dependency measure, as implemented in [9], in which case fitness of each particle was measured by calculating dependency of decision attribute on set of conditional attributes, represented by this particle.

However, we have slightly modified the fitness function and have replaced rough set-based dependency measure with relative attribute dependency. The reason behind is that relative dependency measure intends to avoid the computationally expensive positive region, thus increasing computational efficiency of algorithm. Any other feature selection algorithm can be used here.

5.7.2 Using Relative Dependency Algorithm to Optimize the Selected Features

Preprocessor step provides initial reduct set. However, as inherent with heuristic-based algorithms, the selected attributes may have many irrelevant attributes. We have used the relative dependency algorithm to optimize the generated reduct set. Applying the relative dependency algorithm at this stage however does not degrade the performance as the input set for the algorithm has already been reduced by preprocessor. Algorithm evaluates the reduct set in conventional way to find out if there is any irrelevant feature.

Using hybrid approach provides us with the positive features of both approaches. The following are some advantages of the proposed hybrid approach:

- Using heuristic algorithm lets us produce the candidate reduct set without drowning deep in exhaustive search, which results in cutting down of the execution time and thus makes use of the proposed approach possible for average and large datasets.

- Using relative dependency as optimizer lets us avoid the calculation of computationally expensive positive region which prohibits the use of conventional rough set-based dependency measure for feature selection in datasets beyond smaller size.
- Using relative dependency algorithm, after preprocessor, lets us further optimize the generated reduct set. So, it will be the set of minimum possible attributes that can be considered as final required reduct set. An important point to note here is that relative dependency algorithm at this stage does not hurt the execution time as the input dataset has already been squeezed by Step 1.

The proposed solution provides a unique and novel way to perform feature selection. It differs from tradition feature selection algorithms by extracting the positive features of both exhaustive and heuristic-based algorithms instead of using them as stand-alone. Both of these search categories discussed in Sect. 5.1 have strong and weak points, e.g. exhaustive search is best to find the optimal solution as compared to other categories; however, it requires lot of resources, and in case of larger problem space, it becomes totally unpractical. So, for feature selection in larger datasets, exhaustive search is not a good option. The proposed solution, unlike the other algorithms, uses its strength to optimize the final result instead of using it directly on the entire dataset. So instead of using entire set of attributes as input for exhaustive search algorithm, it is given the reduced set which is already a candidate solution. So, unlike other algorithms, the role of exhaustive search is reduced to optimize the already found solution instead of finding it.

On the other hand, heuristic-based search does not dig deep into problem space to try every solution. The ultimate consequence is that it does not ensure the optimal result set, e.g. in the case of using GA for feature selection, it can find the best chromosome in the first n iterations; however, does this chromosome represent the minimal feature set? You cannot ensure. However, this feature enables heuristic-based approaches to be used for larger datasets. The proposed approach, in contrast to other feature selection approaches, uses heuristic-based algorithms to find the initial feature set rather than the final solution. So, the role of heuristic-based algorithm is limited to find the initial solution instead of the final one, which enables the proposed solution best option for larger datasets.

With the above facts, it can be said that proposed hybrid approach is suitable for feature selection in large datasets to find optimal solution. Experimental results have justified the proposed approach, both in terms of efficiency and effectiveness.

5.8 A Rough Set-Based Feature Selection Approach Using Random Feature Vectors

In [10] authors have proposed a new feature selection method using random feature vectors. A random feature vector contains attributes that are randomly selected. Attribute can be selected by any mechanism. Authors, however, have proposed that each attribute should have 50% probability to be selected. This will give each attribute an equal opportunity to participate in feature subset selection. Algorithm comprises two steps: Firstly it constructs random feature vector. Algorithm first constructs a number of random feature vectors and selects one to be a feature subset. Secondly it optimizes the selected feature vector to eliminate the redundant features. Now we explain both of these steps in detail.

Step 1 Initially algorithm initializes a feature vector by randomly selecting features from dataset. Features are selected on the basis of equal opportunity, i.e. each feature has 50% probability of being selected. This is done by using a random function. The presence of an attribute in the feature vector is marked by its serial number in the dataset; absence on the other hand is marked by placing a zero (0) in its place. So the size of the feature vector is equal to the total number of attributes in dataset; however, some locations of the feature vector will be occupied by attributes and some will be empty, marked by zero. For example, if a dataset has five attributes {a, b, c, d, e}, the random feature vector V = {0, 2, 0, 4, 0} means that the second and fourth attributes will be part of feature vector, while the first, third and fifth attribute will not be present.

After selecting feature vector, algorithm checks its fitness for being selected as a feature subset. Fitness comprises checking dependency of decision class on set of conditional attributes represented by random feature vector. It should be equal to that of the entire set of conditional attributes. If the current feature vector does not fulfil the criteria, algorithm constructs another random feature vector and checks its fitness. Process continues until we find a feature vector which fulfils the specified criteria.

Here, it should be noted that we may not find a feature subset after a number of attempts, in which case, we can increase the probability of selecting an attribute to be part of random feature vector. Once a required feature vector is selected, algorithm proceeds with the second step, i.e. to optimize it.

Step 2 In the second step, algorithm optimizes the selected feature subset. For this purpose, it considers removal of each attribute one by one and checks dependency of remaining attributes. If removal of attribute does not affect dependency of decision class on remaining attributes, this attribute can safely be removed; otherwise, removal of the next attribute is considered. For example, if feature vector V = {1,0,3,0,5} is selected, then algorithm will first remove the first attribute and will check dependency of decision class on the remaining attribute set comprising the third and fifth attribute. If removal of first attribute does not affect dependency on remaining attributes, first attribute will be removed and then removal of third attribute will be considered. The process will continue until no further attribute can be removed without affecting dependency of the remaining attribute set.

The first step selects feature subset and the second step optimizes it. Selecting features randomly reduces the execution time of the algorithm.

Figure 5.14 shows the pseudocode of the proposed algorithm.

Figure 5.15 shows the diagrammatic representation of the proposed solution.

Fig. 5.14 Proposed algorithm

C: C1,C2,…..Cn set of conditional attributes
D: Decision attribute
(a) do
(b) Initialize $V_i \leftarrow \{X_1, X_2, X_3 \ldots\ldots X_n\}$
(c) $Until\ \gamma(V) = \gamma(C)$
(d) $\forall X \in X_v$
(f) $if\ \gamma(V) = \gamma(V - \{X_v\})$
(g) $V = V - \{X_v\}$
(h) Output V

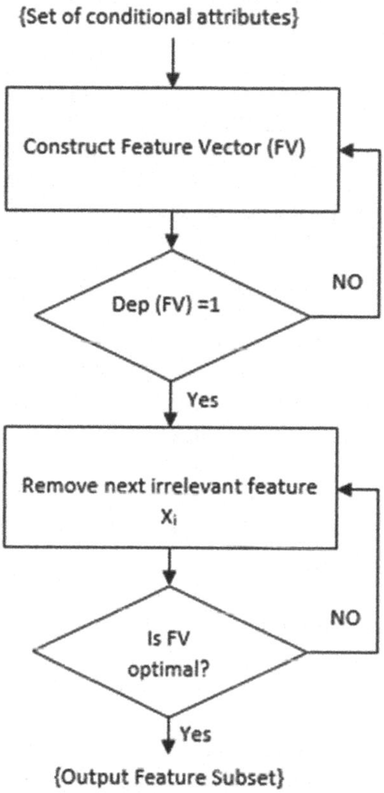

Fig. 5.15 Flow chart of RFS algorithm

Now we explain the algorithm with the help of an example. Consider the dataset given in Table 5.2.

It comprises a decision system having six objects, five conditional attributes C = {a, b, c, d, e} and a decision attribute D = {Z}.

Table 5.2 Sample decision system

U	a	b	c	d	e	Z
1	1	1	2	1	1	1
2	2	2	3	3	2	3
3	1	2	3	2	3	2
4	1	1	1	1	1	1
5	2	3	1	2	2	3
6	2	3	2	3	3	2

The first algorithm calculates dependency of decision class on the entire set of conditional attributes. In this case,

$$\gamma(C) = 1.0$$

For simplicity, we suppose that algorithm constructs feature vector as follows:

V = {1, 2, 0,3,0}, i.e. the first, second and fourth attributes are part of random feature vector. Now algorithm will check dependency of decision class on this feature vector. Here:

$$\gamma(V) = 1.0$$

It means that this feature vector fulfil the criteria and can be selected as feature vector. With this, the first step of algorithm completes. Although the feature subset, i.e. reduct, has been found, it may not be optimal at this stage and may contain redundant attributes. So far,

V = {a,b,d}

Now in the second step, algorithm optimizes it by removing attributes one by one.

The first algorithm removes attribute "a" and checks dependency on the remaining reduct set. It is found that

$$\gamma(V - \{a\}) = 1.0$$

Thus, we conclude that attribute "a" is redundant as its removal does not affect dependency of the remaining attribute set, so now.

V = {b,d}

Now we consider removing the next attribute, i.e. "b", and calculate dependency on the remaining reduct set:

$$\gamma(V - \{b\}) = 0.2$$

Here removing the attribute "b" affects dependency of remaining attribute set, so we cannot remove it. So.

V = {b, d}

Next we consider removal of attribute "d" and calculate dependency on the remaining attribute set:

$$\gamma(V - \{d\}) = 0.2$$

Again we cannot remove attribute "d" without affecting dependency of the remaining reduct set, so reduct set will comprise two attributes, i.e. R = {b, d} which will be an output by algorithm. The proposed solution has two advantages:

First it provides optimal feature subset; there are various feature selection approaches that are based on randomness, e.g. genetic algorithm, particle swarm optimization, fish swarm optimization, etc., but all of these do not ensure the optimal feature subset, and thus the selected feature subset may contain redundant features. However, the proposed solution explicitly removes the redundant features; thus, the output reduct set contains minimum number of features.

Second it does not use any complex operator, e.g. crossover, mutation, changing velocity, etc., thus enhancing overall performance of the algorithm.

Table 5.3 shows the summary of all the rough set-based approaches discussed so far.

5.9 Summary

In this section, we have presented feature selection algorithms using rough set-based positive region and alternate ones. Positive region-based approaches use conventional dependency measure comprising three steps to measure the fitness of an attribute being selected for reduct set. However, using positive region is a computationally expensive approach that makes these approaches inappropriate to use for larger datasets. Alternate approaches are the one that don't use positive region. However, application of such approaches has only been tested against smaller datasets which raises question for their appropriateness for larger datasets.

Table 5.3 Related algorithms based on RST

Algorithm	Technique used	Advantages	Disadvantages
Supervised hybrid feature selection based on PSO and rough sets for medical diagnosis [3]	Particle swarm optimization and rough set-based dependency measure	PSO is an advance heuristic-based algorithm to avoid exhaustive search	Conventional dependency-based measure is a performance bottleneck
Rough set-based genetic algorithm [4]	Conventional positive region-based dependency calculation	It is based on randomness, so the procedure may find reducts in few attempts	Uses conventional positive region-based dependency measure
QuickReduct approach for feature selection [1]	Rough set-based dependency measure	Attempts to calculate reducts without exhaustively generating all possible subsets	Uses conventional dependency-based measure, which is a time-consuming task
ReverseReduct [1].	Rough set-based dependency measure	Backward elimination is utilized without exhaustively generating all possible combinations	Dependency is calculated using conventional positive region-based approach
An incremental algorithm to feature selection in decision systems with the variation of feature set [5]	Incremental feature selection using rough set-based significance measure of attributes	Presents feature selection for dynamic systems where the datasets keep on increasing with time	Conventional dependency measure is used to measure attribute significance. Measuring significance requires measuring dependency twice, once with attribute and then without attribute
Fish swarm algorithm [6]	Rough set-based dependency used with fish swarm method for feature selection	Attempts to find rough set reducts using the swarm logic where swarms can discover best combination of features	Conventional dependency-based measure is again a performance bottleneck
Rough set improved harmony search QuickReduct [7]	Rough set-based dependency measure used with harmony search algorithm for feature selection	Integrates rough set theory with "improved harmony search"-based algorithm with QuickReduct for feature selection	Uses conventional dependency-based measure, which is time-consuming

Bibliography

1. Jensen R, Shen Q (2008) Computational intelligence and feature selection: rough and fuzzy approaches, vol 8. Wiley, Hoboken
2. Pethalakshmi A, Thangavel K (2007) Performance analysis of accelerated quickreduct algorithm. Conference on Computational Intelligence and Multimedia Applications, 2007. International Conference on, vol 2. IEEE
3. Inbarani HH, Azar AT, Jothi G (2014) Supervised hybrid feature selection based on PSO and rough sets for medical diagnosis. Comput Methods Prog Biomed 113(1):175–185
4. Zuhtuogullari K, Allahverdi N, Arikan N (2013) Genetic algorithm and rough sets based hybrid approach for reduction of the input attributes in medical systems. Int J Innov Comput Inf Control 9:3015–3037
5. Qian W et al (2015) An incremental algorithm to feature selection in decision systems with the variation of feature set. Chin J Electron 24(1):128–133
6. Chen Y, Zhu Q, Huarong X (2015) Finding rough set reducts with fish swarm algorithm. Knowl-Based Syst 81:22–29
7. Inbarani H, Hannah MB, Azar AT (2015) A novel hybrid feature selection method based on rough set and improved harmony search. Neural Comput & Applic 26(8):1859–1880
8. Raza MS, Qamar U (2016) A hybrid feature selection approach based on heuristic and exhaustive algorithms using Rough set theory. Proceedings of the International Conference on Internet of things and Cloud Computing. ACM
9. Raza MS, Qamar U (2017) Feature selection using rough set based heuristic dependency calculation. PhD dissertation, NUST
10. Raza MS, Qamar U (2016) A rough set based feature selection approach using random feature vectors. Frontiers of Information Technology (FIT), 2016 International Conference on. IEEE

Chapter 6
Unsupervised Feature Selection Using RST

Supervised feature selection evaluates the features that provide maximum information based on classification accuracy. This requires labelled data; however, in real world not all the data is properly labelled, so we may come across the situation where little or no class information is provided. For such type of data, we need unsupervised feature selection information that could find feature subsets without given any class labels. In this section, we will discuss some of the unsupervised feature subset algorithms based on rough set theory.

6.1 Unsupervised QuickReduct Algorithm (USQR)

In [1] authors have presented an unsupervised QuickReduct algorithm using RST. Original QuickReduct algorithm, which is supervised, takes two inputs, i.e. set of conditional attributes and decision attribute(s). However, USQR takes only one input, i.e. set of conditional attributes. However, just like existing QuickReduct algorithm, it performs feature selection without exhaustively generating all possible subsets.

It starts with an empty set and adds those attributes one by one which result in maximum degree of increase in dependency until it produces maximum possible value. According to the algorithm, the mean dependency of each attribute subset is calculated and the best candidate is chosen:

$$\gamma_P(a) = \frac{|\,\mathrm{POS}_p(a)\,|}{|\,U\,|}, \quad \forall a \in A.$$

The following is the pseudocode of the algorithm (Fig. 6.1):

Now we explain USQR with an example taken from [1]. We will consider the following dataset (Table 6.1).

M.S. Raza, U. Qamar, *Understanding and Using Rough Set Based Feature Selection: Concepts, Techniques and Applications*,
DOI 10.1007/978-981-10-4965-1_6

Fig. 6.1 Unsupervised QuickReduct algorithm

USQR(C)
C, the set of all conditional features;
(1) $R \leftarrow \{\}$
(2) do
(3) $T \leftarrow R$
(4) $\forall x \in (C - R)$
(5) $\forall y \in C$
(6) $\gamma_{RU(x)}(y) = \frac{|POS_{RU(x)}(y)|}{|U|}$
(7) if $\overline{\gamma_{RU(x)}(y), \forall y \in C} > \overline{\gamma_r(y), \forall y \in C}$
(8) $T \leftarrow RU\{x\}$
(9) $R \leftarrow T$
(10) until $\overline{\gamma_R(y), \forall y \in C} = \overline{\gamma_C(y), \forall y \in C}$
return R

Table 6.1 Sample dataset taken from [1]

$x \in U$	a	b	c	d
1	1	0	2	1
2	1	0	2	0
3	1	2	0	0
4	1	2	2	1
5	2	1	0	0
6	2	1	1	0
7	2	1	2	1

Dataset comprises four conditional attribute, i.e. a, b, c and d. In Step 1 we calculate dependency value of each attribute.

Step 1

$$\gamma_{\{a\}}(\{a\}) = \frac{|POS_{\{a\}}(\{a\})|}{|U|} = \frac{|\{1,2,3,4,5,6,7\}|}{|\{1,2,3,4,5,6,7\}|} = \frac{7}{7}$$

$$\gamma_{\{b\}}(\{b\}) = \frac{|POS_{\{a\}}(\{b\})|}{|U|} = \frac{|\{5,6,7\}|}{|\{1,2,3,4,5,6,7\}|} = \frac{3}{7}$$

$$\gamma_{\{c\}}(\{c\}) = \frac{|POS_{\{a\}}(\{c\})|}{|U|} = \frac{|\{\}|}{|\{1,2,3,4,5,6,7\}|} = \frac{0}{7}$$

$$\gamma_{\{d\}}(\{d\}) = \frac{|POS_{\{a\}}(\{d\})|}{|U|} = \frac{|\{\}|}{|\{1,2,3,4,5,6,7\}|} = \frac{0}{7}$$

$$\sum_{\forall y \in c} \gamma_{\{a\}}(\{y\}) = \frac{7}{7} + \frac{3}{7} + \frac{0}{7} + \frac{0}{7} = \frac{10}{7}$$

$$\overline{\gamma_{\{a\}}(\{y\}),\ \forall y \in C} = \frac{\frac{10}{7}}{4} = 0.35714$$

Table 6.2 Degree of dependency after Step 1

y/x	$\{a\}$	$\{b\}$	$\{c\}$	$\{d\}$
a	1.0000	1.0000	0.1429	0.0000
b	0.4286	1.0000	0.1429	0.0000
c	0.0000	0.2857	1.0000	0.4286
d	0.0000	0.0000	0.4286	1.0000
$\overline{\gamma_{\{a\}}(\{y\}), \forall y \in C}$	0.3571	0.3571	0.3571	0.3571

Similarly, other degrees of dependency values are calculated. Table 6.2 shows these values.

Attribute b generates the highest degree of dependency, hence chosen to evaluate the indiscernibility of sets {a, b}, {b, c} and {b, d} and calculate the degree of dependency as given in Step 2.

Step 2

$$\gamma_{\{a,b\}}(\{a\}) = \frac{|POS_{\{a,b\}}(\{a\})|}{|U|} = \frac{|\{1,2,3,4,5,6,7\}|}{|\{1,2,3,4,5,6,7\}|} = \frac{7}{7}$$

$$\gamma_{\{a,b\}}(\{b\}) = \frac{|POS_{\{a,b\}}(\{b\})|}{|U|} = \frac{|\{1,2,3,4,5,6,7\}|}{|\{1,2,3,4,5,6,7\}|} = \frac{7}{7}$$

$$\gamma_{\{a,b\}}(\{c\}) = \frac{|POS_{\{a,b\}}(\{c\})|}{|U|} = \frac{|\{1,2\}|}{|\{1,2,3,4,5,6,7\}|} = \frac{2}{7}$$

$$\gamma_{\{a,b\}}(\{d\}) = \frac{|POS_{\{a,b\}}(\{d\})|}{|U|} = \frac{|\{\}|}{|\{1,2,3,4,5,6,7\}|} = \frac{0}{7}$$

$$\sum_{\forall y \in c} \gamma_{\{a,b\}}(\{y\}) = \frac{7}{7} + \frac{7}{7} + \frac{2}{7} + \frac{0}{7} = \frac{16}{7}$$

$$\overline{\gamma_{\{a,b\}}(\{y\}),\ \forall y \in C} = \frac{\frac{16}{7}}{4} = 0.57143$$

Similarly the other dependency values are shown in Table 6.3.

Subsets {b,c} and {b,d} generate the highest degree of dependency, but algorithm selects {b,c} as it appears first and calculates indiscernibility of {a,b,c} and {b,c,d} as shown in Step 3.

Step 3

$$\gamma_{\{a,b\}}(\{a\}) = \frac{|POS_{\{a,b,c\}}(\{a\})|}{|U|} = \frac{|\{1,2,3,4,5,6,7\}|}{|\{1,2,3,4,5,6,7\}|} = \frac{7}{7}$$

Table 6.3 Degree of dependency after Step 2

y/x	$\{a, b\}$	$\{b, c\}$	$\{b, d\}$
a	1.0000	1.0000	1.0000
b	1.0000	1.0000	1.0000
c	0.2857	1.0000	0.7143
d	0.0000	0.7143	1.0000
$\overline{\gamma_{\{a\}}(\{y\}), \forall y \in C}$	0.57143	0.57143	0.57143

$$\gamma_{\{a,b,c\}}(\{b\}) = \frac{|POS_{\{a,b,c\}}(\{b\})|}{|U|} = \frac{|\{1,2,3,4,5,6,7\}|}{|\{1,2,3,4,5,6,7\}|} = \frac{7}{7}$$

$$\gamma_{\{a,b,c\}}(\{c\}) = \frac{|POS_{\{a,b,c\}}(\{c\})|}{|U|} = \frac{|\{1,2\}|}{|\{1,2,3,4,5,6,7\}|} = \frac{2}{7}$$

$$\gamma_{\{a,b,c\}}(\{d\}) = \frac{|POS_{\{a,b,c\}}(\{d\})|}{|U|} = \frac{|\{3,4,5,6,7\}|}{|\{1,2,3,4,5,6,7\}|} = \frac{5}{7}$$

$$\sum_{\forall y \in c} \gamma_{\{a,b,c\}}(\{y\}) = \frac{7}{7} + \frac{7}{7} + \frac{7}{7} + \frac{5}{7} = \frac{26}{7}$$

$$\overline{\gamma_{\{a,b,c\}}(\{y\}), \forall y \in C} = \frac{\frac{16}{7}}{4} = 0.02857$$

Similarly:

$$\overline{\gamma_{\{b,c,d\}}(\{y\}), \forall y \in C} = 1$$

The other dependency values calculated in Step 3 are shown in Table 6.4.

Since the dependency value of subset {b,c,d} is one, the algorithm terminates and outputs this subset as reduct.

6.2 Unsupervised Relative Reduct Algorithm

In [2] authors have presented a relative dependency-based algorithm for unsupervised datasets (US relative dependency). Existing relative dependency algorithm uses both conditional and decision attributes for feature subset selection. However, US relative dependency performs feature selection on the basis of relative dependency using only the conditional set of attributes.

Figure 6.2 shows the pseudocode of the algorithm.

Initial feature subset comprises all the features in dataset. It then evaluates each feature. If relative dependency of the feature is “1”, it can safely be removed. The relative dependency for unsupervised data can be calculated as follows:

Table 6.4 Degree of dependency after Step 4

y/x	$\{a, b, c\}$	$\{b, c, d\}$
a	1.0000	1.0000
b	1.0000	1.0000
c	1.0000	1.0000
d	0.7143	1.0000
$\gamma_{\{a\}}(\{y\}), \forall y \in C$	0.9285	1.0000

Fig. 6.2 Unsupervised QuickReduct algorithm

```
USRelativeReduct(C)
C, the conditional attributes;
    (1) R← C
    (2) ∀ a ∈ C
    (3) if(K_{R-{a}}({a}) == 1)
    (4) R ← R − {a}
return R
```

$$K_R(\{a\}) = \frac{|\,U/IND(R)\,|}{|\,U/IND(R \cup \{a\})\,|}, \quad \forall a \in A$$

Then show that R is a reduct if and only if $K_R(\{a\}) = K_C(\{a\})$ and $\forall X \subset R$, $K_X(\{a\}) \neq K_C(\{a\})$. In this case, the decision attribute used in the supervised feature selection is replaced by the conditional attribute a, which is to be eliminated from the current reduct set R.

Now we will explain unsupervised relative reduct with an example taken from [2]. We will consider the following dataset (Table 6.5):

As the algorithm uses backward elimination, initially reduct set comprises the entire set of conditional attributes, i.e. R = {a,b,c,d}. Now the algorithm considers attribute "a" for elimination:

$$K_{\{b,c,d\}}(\{a\}) = \frac{\left|\frac{U}{IND(b,c,d)}\right|}{\left|\frac{U}{IND(a,b,c,d)}\right|} = \frac{|\{\{1\}\{2\}\{3\}\{4\}\{5\}\{6\}\{7\}\}|}{|\{\{1\}\{2\}\{3\}\{4\}\{5\}\{6\}\{7\}\}|} = \frac{7}{7}$$

As dependency is equal to "1", so attribute "a" can be safely removed. The reduct set thus becomes R = {b,c,d}.

Now we consider elimination of "b":

$$K_{\{c,d\}}(\{b\}) = \frac{\left|\frac{U}{IND(c,d)}\right|}{\left|\frac{U}{IND(b,c,d)}\right|} = \frac{|\{\{1,4,7\}\{2\}\{3,5\}\{6\}\}|}{|\{\{1\}\{2\}\{3\}\{4\}\{5\}\{6\}\{7\}\}|} = \frac{4}{7}$$

As dependency is not equal to "1", we cannot remove "b". The next algorithm considers elimination of attribute "c":

Table 6.5 Sample dataset taken from [2]

$x \in U$	a	b	c	d
1	1	0	2	1
2	1	0	2	0
3	1	2	0	0
4	1	2	2	1
5	2	1	0	0
6	2	1	1	0
7	2	1	2	1

$$K_{\{b,d\}}(\{c\}) = \frac{\left|\frac{U}{IND(b,d)}\right|}{\left|\frac{U}{IND(b,c,d)}\right|} = \frac{|\{\{1\}\{2\}\{3\}\{4\}\{5,6\}\{7\}\}|}{|\{\{1\}\{2\}\{3\}\{4\}\{5\}\{6\}\{7\}\}|} = \frac{6}{7}$$

Again relative dependency does not evaluate to "1" so we cannot eliminate "*c*" as well. Now algorithm considers elimination of attribute "*d*":

$$K_{\{b,c\}}(\{d\}) = \frac{\left|\frac{U}{IND(b,c)}\right|}{\left|\frac{U}{IND(b,c,d)}\right|} = \frac{|\{\{1,2\}\{3\}\{4\}\{5\}\{6\}\{7\}\}|}{|\{\{1\}\{2\}\{3\}\{4\}\{5\}\{6\}\{7\}\}|} = \frac{6}{7}$$

Again relative dependency is not equal to "1", so we cannot remove "*d*". Thus reduct set comprises attribute R = {b,c,d}.

6.3 Unsupervised Fuzzy-Rough Feature Selection

In [3] authors have presented an unsupervised fuzzy-rough feature selection algorithm with different fuzzy-rough feature evaluation criteria. Algorithm starts by considering all the features in the dataset. It then evaluates the measure used as evaluation criteria without this feature. If the measure remains unaffected, then feature is removed. The process continues until no further features can be removed without affecting the corresponding measure.

Following evaluation measures were used:

Dependency Measure A set of attribute(s) Q depends on set of attribute(s) P if P uniquely determines Q. However, the authors argue that fuzzy dependency measure along with its use for supervised fuzzy-rough feature selection can also be used for determining the interdependencies between attributes. This can be achieved by replacing the decision class with the set of feature Q.

Boundary Region Measure Most of the approaches in crisp-rough set-based feature selection and all the approaches of fuzzy-rough feature selection use lower approximation for feature selection; however, upper approximation can

Fig. 6.3 Unsupervised QuickReduct algorithm

```
UFRQUICKREDUCT(F)
F, the set of all features.
    (1) R ← ℂ
    (2) foreach x ∈ ℂ
    (3) R ← R − {x}
    (4) if M(R, {x}) < 1
    (5) R ← R U {x}
    (6) return R
```

also be used to discriminate between objects. For example, two subsets may produce the same lower approximation, but one may give smaller upper approximation, meaning the lesser uncertainty in boundary region. Similarly fuzzy boundary region can be used for feature evaluation.

Discernibility Measure In the case of conventional RST, the feature selection algorithms can be categorized in two broad classes: those using dependency measure and those using discernibility measure. The fuzzy tolerance relations that represent objects' approximate equality can be used to extend the classical discernibility function. For each combination of features P, a value is obtained indicating how well these attributes maintain the discernibility, relative to another subset of features Q, between all objects.

The pseudocode of the algorithm is given in Fig. 6.3.

Algorithm can be used by specifying any of the evaluation measure mentioned above. The complexity for the search in the worst case is O(n), where n is the number of original features.

6.4 Unsupervised PSO-Based Relative Reduct (US-PSO-RR)

Authors in [4] have presented a hybrid approach for unsupervised feature selection. The approach uses relative reduct and particle swarm optimization for this purpose. Figure 6.4 shows the pseudocode of their proposed algorithm. Algorithm takes set of conditional attributes "C" as input and produces the reduct set "R" as output.

In Step 1, the proposed algorithm initializes the initial populations of particles with randomly selected conditional attributes and initial velocity. A population of particles is constructed then. For each particle, mean relative dependency is calculated. If the dependency is "1", it is considered as reduct set. If mean dependency is not equal to "1", then the Pbest (highest relative dependency value) of each particle is retained, and the best value of the entire population is retained as the globalbest value. Algorithm finally updates position and velocity of the next population generated.

```
Algorithm:   US-PSO-RR(C)
Input: C, the set of all conditional features,
Output: Reduct R

Step 1: Initialize X with random position Vi with random velocity
∀: Xi ← randomPosition();
Vi ← randomVelocity();
fit ← 0; globalbest ← fit;
gbest ← X1; pbest(1) ← X1
For i = 1 ... S
pbest(i) = Xi
Fitness(i) = 0
End for

Step 2: While Fitness != 1  // stopping criterion
For i = 1 ... S  // for each particle
∀: Xi;//Compute fitness of feature subset of Xi
R ← Feature subset of Xi (1's of Xi)
∀ a ∈ (y)
γR(a) = |U/IND(R)| / |U/IND(RU{a})|
Fit = γR(y)∀ y ∉ R
if Fitness(i) > fit
Fitness(i) = fit
pbest(i) = Xi
End
if(Fit == 1)
return R
End if
End for

Step 3: Compute best fitness
For i = 1, ...,S
if(Fitness(i) > globalbest)
gbest ← Xi ;
globalbest ←Fitness(i);
End if
End for
UpdateVelocity(); // Update velocity Vi's of Xi's
UpdatePosition(); // Update position of Xi's, Continue with the next iteration
End {while}
Output Reduct R
```

Fig. 6.4 Pseudocode of US-PSO-RR

Encoding:

Algorithm uses "1" and "0" to represent presence and absence of attributes. The attribute that will be included as part of particle position is represented by "1", and the particle which will not be part of the particle will be represented by "0". For example, if there are five attributes, i.e. a, b, c, d and e, and we need to include the attributes b, c and e, the particle will be like

a	b	c	d	e
0	1	1	0	1

The above particle position shows that the particle will include attributes b, c, and e, whereas attributes a and d will be absent.

Representation and Updates of Velocity and Positions.

Velocity of particles is represented by positive integer between 1 and $V_{max.}$ It basically determines how many bits of the particle should be changed with respect to globalbest position. Pbest represents the localbest and gbest represents the globalbest index. The velocity of each particle is updated according to the following equation:

$$V_{id} = w^* V_{id} + c_1{}^* rand()^* (P_{id} - x_{id}) + c_2{}^* Rand()^* \left(P_{gd} - x_{id}\right)$$

Here w represents inertia weight and c1 and c2 represent acceleration constants. Particle's position is changed on the basis of velocity as follows:

- If V ≤xg, randomly change V bits of the particle, which are different from that of gbest.
- If V >xg, change all the different bits to be the same as that of gbest, and further (V−xg) bits should be changed randomly.

W can be calculated as follows:

$$w = w_{max} - \frac{w_{max} - w_{min}}{iter_{max}} \text{ iter}$$

Here w_{max} is the initial value of the weighting coefficient, w_{min} is the final value, $iter_{max}$ is the maximum number of iterations, and iter is the current iteration.

Algorithm uses relative dependency measure to measure the fitness of a particle. Relative dependency is calculated as follows:

$$\gamma_R(a) = \left| \frac{U/IND(R)|}{|U/IND(R \cup \{a\})| \forall a \notin R} \right.$$

where R is the subset selected by the particle, and the mean dependency of selected gene subset, on all the genes that are not selected by the particle, is used as the fitness value of the particle X_i.

$$\text{Fitness} = Fitness(X_i) = \overline{\gamma_R}(Y) \forall y \notin R$$

We will now explain this algorithm with the help of an example taken from [5]. We will use the following dataset as sample:

Suppose initially the following particle generated was (1,0,0,1). This particle contains the features "a" and "d" and excludes features "b" and "c". Hence R = {a, d} and Y = {b,c}.

Now,

$$\gamma_R(b) = \frac{|INDR_R|}{|INDR_{R\cup\{b\}}|} = \frac{|\{1,4\}\{2,3\}\{5,6\}\{7\}|}{|\{1\}\{2\}\{3\}\{4\}\{5\}\{6\}\{7\}|} = \frac{4}{6} = 0.667$$

$$\gamma_R(c) = \frac{|INDR_R|}{|INDR_{R\cup\{c\}}|} = \frac{|\{1,4\}\{2,3\}\{5,6\}\{7\}|}{|\{1\}\{2\}\{3\}\{4\}\{5\}\{6\}\{7\}|} = \frac{4}{6} = 0.667$$

$$\overline{\gamma_R}(a)\forall a \in Y = \frac{0.667 + 0.667}{2} = 0.667$$

Since $\overline{\gamma_R}(a) \neq 1$, so {a,d} will not be considered as reduct set. Suppose at some stage a particle Xi = {0,1,1,1} is generated. Dependency will be as follows:
R = {b,c,d} and Y = {a}

$$\gamma_R(a) = \frac{|INDR_R|}{|INDR_{R\cup\{a\}}|} = \frac{|\{1\}\{2\}\{3\}\{4\}\{5\}\{6\}\{7\}|}{|\{1\}\{2\}\{3\}\{4\}\{5\}\{6\}\{7\}|} = \frac{7}{7} = 1$$

$$\overline{\gamma_R}(a)\forall a \in Y = 1$$

So R = {b,c,d} will be the reduct set.

6.5 Unsupervised PSO-Based Quick Reduct (US-PSO-QR)

US-PSO-QR works use the same mechanism as discussed in US-PSO-QR. Figure 6.5 shows pseudocode of the algorithm.

Algorithm takes set of conditional attributes as input and produces reduct set "R" as output. It starts with initial population of particles and evaluates fitness of each particle. A feature with highest fitness is selected and its combinations with other features are evaluated. The local- and globalbest (pbest and gbest) are updated accordingly. Finally algorithm updates velocity and position of each particle. The process continues until we meet the stopping criteria which normally comprises the maximum number of iterations.

Now we will explain this algorithm with the help of example. We will consider Table 6.6 and the same initial population.

$$\gamma_{T\cup\{ad\}}(a) = \frac{|POS_{T\cup\{ad\}}(a)|}{|U|} = \frac{|\{1,2,3,4,5,6,7\}|}{|\{1,2,3,4,5,6,7\}|} = \frac{7}{7} = 1$$

$$\gamma_{T\cup\{ad\}}(b) = \frac{|POS_{T\cup\{ad\}}(b)|}{|U|} = \frac{|\{5,6,7\}|}{|\{1,2,3,4,5,6,7\}|} = \frac{3}{7} = 0.4286$$

$$\gamma_{T\cup\{ad\}}(c) = \frac{|POS_{T\cup\{ad\}}(c)|}{|U|} = \frac{|\{1,4,7\}|}{|\{1,2,3,4,5,6,7\}|} = \frac{3}{7} = 0.4286$$

Algorithm: US-PSO-QR(C)
Input: C, the set of features,
Output: Reductset R

Step 1: Initialize X with random position V_i with random velocity
$\forall: X_i \leftarrow randomPosition();$
$V_i \leftarrow randomVelocity();$
fit ← 0; globalbest ← Fit;
gbest ← X_1;

Step 2: While Fitness != $\overline{\gamma_C}(y)\ \forall\, y \in C$ // stopping criterion
For i = 1 … S // for each particle
$\forall: X_i;$ T ← {}
//Compute fitness of feature subset of X_i
R ← Feature subset of X_i (1's of X_i)
$\forall\, x \in R;\ \forall\, y \in C$
$\gamma_{TU(x)}(y) = \frac{|POS_{TU(x)}(y)|}{|U|}$
$Fitness(i) = \overline{\gamma_{TU(x)}}(y)\ \forall\, y \in C$
End for
Step 3: Compute best fitness
For i = 1, …,S
if(fitness(i) > globalbest)
gbest ← X_i; globalbest ← Fitness(i); pbest(i) ← bestPos(X_i);
$if\ fitness(i) = \overline{\gamma_C}(y)\ \forall\, y \in C$
R ← getReduct(X_i)
End if
End if
End for
UpdateVelocity(); // Update velocity V_i's of X_i's
UpdatePosition(); // Update position of X_i's, Continue with the next iteration
End {while}
Output Reduct R

Fig. 6.5 Pseudocode of US-PSO-RR

$$\gamma_{T\cup\{ad\}}(d) = \frac{|POS_{T\cup\{ad\}}(d)|}{|U|} = \frac{|\{1,2,3,4,5,6,7\}|}{|\{1,2,3,4,5,6,7\}|} = \frac{7}{7} = 1$$

$$\overline{\gamma_{T\cup\{ad\}}}(y); \forall y \in C = \frac{1 + 0.4286 + 0.4286 + 1}{4} = 0.7143$$

$$\overline{\gamma_{T\cup\{bd\}}}(y); \forall y \in C = \frac{|POS_{T\cup\{bd\}}(d)|}{|U|} = 0.9286$$

$$\overline{\gamma_{T\cup\{ab\}}}(y); \forall y \in C = \frac{|POS_{T\cup\{ab\}}(d)|}{|U|} = 0.5714$$

$$\overline{\gamma_{T\cup\{bc\}}}(y); \forall y \in C = \frac{|POS_{T\cup\{bc\}}(d)|}{|U|} = 0.9286$$

Table 6.6 Sample dataset taken from [5]

$x \in U$	a	b	c	d
1	1	0	2	1
2	1	0	2	0
3	1	2	0	0
4	1	2	2	1
5	2	1	0	0
6	2	1	1	0
7	2	1	2	1

Since none of the attribute shows dependency of "1", the next iteration starts. Suppose at any stage the particle is (0,1,1,1), dependency will be calculated as

$$\gamma_{T\cup\{bcd\}}(a) = \frac{|POS_{T\cup\{bcd\}}(a)|}{|U|} = \frac{|\{1,2,3,4,5,6,7\}|}{|\{1,2,3,4,5,6,7\}|} = \frac{7}{7} = 1$$

$$\gamma_{T\cup\{bcd\}}(b) = \frac{|POS_{T\cup\{bcd\}}(b)|}{|U|} = \frac{|\{1,2,3,4,5,6,7\}|}{|\{1,2,3,4,5,6,7\}|} = \frac{7}{7} = 1$$

$$\gamma_{T\cup\{bcd\}}(c) = \frac{|POS_{T\cup\{bcd\}}(c)|}{|U|} = \frac{|\{1,2,3,4,5,6,7\}|}{|\{1,2,3,4,5,6,7\}|} = \frac{7}{7} = 1$$

$$\gamma_{T\cup\{bcd\}}(d) = \frac{|POS_{T\cup\{bcd\}}(d)|}{|U|} = \frac{|\{1,2,3,4,5,6,7\}|}{|\{1,2,3,4,5,6,7\}|} = \frac{7}{7} = 1$$

$$\overline{\gamma_{T\cup\{bcd\}}}(y); \forall y \in C = 1$$

The subset {b,c,d} produces the dependency equal to "1"; hence algorithm stops and outputs R = {b,c,d} as reduct.

6.6 Summary

There are a number of feature selection algorithms employing rough set theory for unsupervised datasets. RST theory is equally effective for unsupervised feature selection as in case of supervised mod. In this chapter we have presented few of the most commonly refereed feature selection algorithms. The details of each algorithm along description of pseudocode were also provided. Working examples are also explained in order to mention the exact work of the algorithms.

Bibliography

1. Velayutham C, Thangavel K (2011) Unsupervised quick reduct algorithm using rough set theory. J Electronic Sci Technol 9.3:193–201
2. Velayutham C, Thangavel K (2011) Rough set based unsupervised feature selection using relative dependency measures. In: Digital proceedings of UGC sponsored national conference on emerging computing paradigms. 2011
3. Parthaláin NM, Jensen R (2010) Measures for unsupervised fuzzy-rough feature selection. Int J Hybrid Intell Syst 7(4):249–259
4. Inbarani HH, Nizar Banu PK (2012) Unsupervised hybrid PSO—relative reduct approach for feature reduction. In: Pattern Recognition, Informatics and Medical Engineering (PRIME), International Conference on. IEEE, 2012
5. Nizar Banu PK, Hannah Inbarani H (2012) Performance evaluation of hybridized rough set based unsupervised approaches for gene selection. Int J Computational Intell Inf 2(2):132–141

Chapter 7
Critical Analysis of Feature Selection Algorithms

So far in previous chapters, we have discussed details of various feature selection algorithms, both rough set based and non-rough set based, for supervised learning and unsupervised learning. In this chapter we will provide analysis of different RST-based feature selection algorithms. With explicit discussion on their results, different experiments were performed to compare the performance of algorithms. We will focus on RST-based feature selection algorithms.

7.1 Pros and Cons of Feature Selection Techniques

Feature selection algorithms can be classified into three categories, as follows:

- Filter methods
- Wrapper methods
- Embedded methods

We have already provided a detailed discussion on all of these techniques. Here we will discuss some plus and negative points of all of these approaches.

7.1.1 Filter Methods

Filter methods perform feature selection independent of the underlying learning algorithm. Normally such methods rank feature according to some ranking criteria and select the best features. They ignore the impact of the feature on learning algorithm. Here are some advantage and disadvantages of this approach:

M.S. Raza, U. Qamar, *Understanding and Using Rough Set Based Feature Selection: Concepts, Techniques and Applications*,
DOI 10.1007/978-981-10-4965-1_7

Pros

- Scalable to high-dimensional datasets
- Simple and fast
- Independent of underlying classification algorithm, so can be used with any classification algorithm

Cons
As classification algorithm is ignored during feature ranking, selection of features may affect accuracy and performance of classification algorithm.

- Ranks features independently and thus ignores dependencies among features.
- Normally univariate or low variate.

7.1.2 Wrapper Methods

Perform feature selection based on classification algorithm. Features are selected by using the feedback from classifier. The following are some of the advantages and disadvantages of this approach:

Pros

- Feature selection is performed by using feedback of the classification algorithm.
- Enhances performance and accuracy of the classifier.
- Feature dependencies are also considered.

Cons

- As compared to filter techniques, wrappers are highly prone to over fitting.
- Computationally expensive.

7.1.3 Embedded Methods

Perform feature selection as part of the learning procedure, e.g. classification trees, learning machines, etc.

Pros

- As compared to wrapper methods, they are less expensive.

Cons

- Learning machine dependent.

7.2 Comparison Framework

A comparison framework was designed by authors in [1] to perform analysis of different feature selection algorithms. The framework comprises three components discussed below:

7.2.1 *Percentage Decrease in Execution Time*

Percentage decrease in execution time specifies the efficiency of an algorithm in terms of how fast it is and how much execution time it cuts down. For this purpose system stopwatch was used, which after feeding the input was started and after getting the results was stopped. The formula to calculate the % decrease is as follows:

$$\text{Percentage decrease} = 100 - \frac{E(1)}{E(2)} * 100,$$

where $E(1)$ is execution time of one algorithm and $E(2)$ is that of its competitor.

7.2.2 *Memory Usage*

Memory usage specifies the maximum amount of runtime memory taken by the algorithm to complete the task taken during its execution. Memory usage metric calculates memory by summing the size of each of the intermediate data structure used.

We have executed feature selection algorithms using "Optidigits" dataset from UCI [2]. The Table 7.1 shows the results obtained (Fig. 7.1).

It can be noted that FSHE took maximum execution time. It is due to the reason that it first uses heuristic approach and then applies exhaustive search on the result produced. The minimum time was taken by genetic algorithm. It gave maximum decrease in execution time w.r.t. FSHE algorithm.

As far as memory is concerned, the same amount of memory was taken by all algorithms as the dataset used was the same. Note that we only considered the major data structures to calculate memory and the intermediate/local variables were neglected.

Table 7.1 Execution time of feature selection algorithms

PSO-QR time (m)	FSA time (m)	RS-IHS-QR time (m)	GA time (m)	IFSA time (m)	FSHE time (m)
6.52	2.21	0.46	0.40	1.22	12.24
46.11%	81.73%	96.19%	96.69%	89.91%	–

Fig. 7.1 shows graph of comparison of execution time

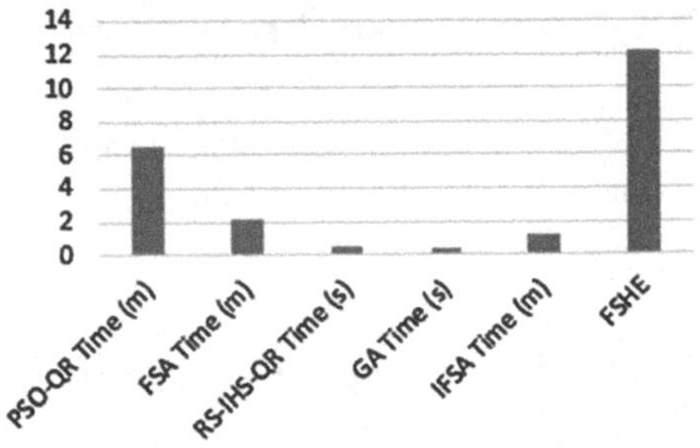

7.3 Critical Analysis of Various Feature Selection Algorithms

Now we will present critical analysis of various feature selection techniques. Strengths and weaknesses of each will be discussed.

7.3.1 *QuickReduct*

QuickReduct [3] attempts to find feature subset without exhaustively generating all possible subsets. It is one of the most commonly referred algorithms; however, QuickReduct does not ensure optimal feature subset selection. During experimentation, it was found that performance of QuickReduct depends on the distribution of attributes, e.g. in the case of forward feature selection, if attributes that result in higher degree of dependency are found at the end, the algorithm will take more time as compared to if the attributes are found in starting indexes. We explain this with the help of the following example. Consider the following dataset (Table 7.2a):

In the first iteration, algorithm selects "b" as part of reduct set; however, in the second iteration, algorithm will pick at the first iteration "a,b" as reduct because the set {a,b} results in dependency equal to "1". However, if attributes are distributed as follows (Table 7.2b):

Then in the second iteration, algorithm will first combine "b" with "a" and then "b" with "c", hence taking more time. Thus performance of algorithm depends on the distribution of attributes. Similarly, the number of attributes in reduct set may also vary depending upon distribution of attributes.

Table 7.2a Sample dataset

U	a	b	c	D
X_1	0	0	0	x
X_2	1	1	1	y
X_3	1	1	0	y
X_4	2	0	1	x
X_5	0	2	0	z
X_6	2	2	0	y

Table 7.2b Sample dataset after changing attribute sequence

U	c	b	a	D
X_1	0	0	0	x
X_2	1	1	1	y
X_3	0	1	1	y
X_4	1	0	2	x
X_5	0	2	0	z
X_6	0	2	2	y

7.3.2 *Rough Set-Based Genetic Algorithm*

In [4] authors present a genetic algorithm for feature subset selection. Genetic algorithms have been successfully applied to find feature subsets; however, these algorithms suffer the following drawbacks:

- Randomness forms the core of genetic algorithms; however, this may lead to lose important features containing more information than others, e.g. in a chromosome, there may be attributes having lower individual dependency but cumulative dependency of "1" and thus the attributes are represented through genes in such chromosome qualifying for the feature subset. However, the dataset may still have attributes having higher dependency (thus more information) as compared to these low dependency attributes (represented by chromosome).
- Genetic algorithms do not ensure the optimal feature subset selection as attributes are randomly selected, so the chromosome may contain redundant attributes, e.g. a chromosome may contain ten attributes, out of which only three may be sufficient (providing dependency equal to that of the entire attribute set).
- Evaluation of fitness function normally acts as bottleneck for genetic algorithm. In context of RST-based feature selection, normally the evaluation function comprises calculating dependency of decision class "D" on the set of conditional attributes encoded into current chromosome. However, for large population size (more number of chromosomes) or larger datasets, calculating dependency is really a computationally expensive job seriously affecting the performance of algorithm.

- Normally algorithm requires more number of chromosomes in a population (larger population size), and in some cases (as observed in experimentation), the algorithm was explicitly forced to stop after specified number of iterations without producing the optimal results.
- The operators like crossover, mutation, etc., are important for the efficiency of the algorithm, but there are no proper heuristics as to use which operator in which situation. Same is the case with chromosome encoding scheme.
- The stopping criteria are not clear, e.g. the maximum number of iterations (in the case of no optimal solution found) is manually specified and has no guidelines regarding it.

7.3.3 PSO-QR

PSO-QR [5] is rough set-based hybrid feature selection approach using swarm optimization combined with QuickReduct. The intention was to take advantage of both QuickReduct and PSO. Algorithm, however, inherits some limitations of both. Here we mention some of the limitations as follows:

- As with genetic algorithm, PSO-QR uses random particle generation, "1" represents the presence of an attribute and "0" represents absence. For example, the following particle shows that the first, third and fifth attributes will be considered and the second and fourth will be missed (Fig. 7.2).

  However, it suffers the same problem, i.e. particles may contain redundant attributes. So again the result produced is not optimal.
- It uses rough set-based dependency measure based on positive region, so not suitable for particles with large size of swarm with large number of particles.
- Furthermore, it uses QuickReduct as shown in the following step of the algorithm, which causes serious performance bottlenecks (Fig. 7.3).
- There are no proper guidelines to select the values of inertia weights; furthermore, there are chances that algorithm may be trapped in the local optima.
- There are no clear stopping criteria as well. Normally algorithm terminates after a specified number of iterations whether the ideal solution has been reached or not.
- The method easily suffers from the partial optimism, which causes the less exact at the regulation of its speed and the direction [6].

It should, however, be noted that PSO does not involve the genetic operators like crossover, mutation, etc. However, particles change their velocity and position with respect to the globalbest particle. It means less change as compared to chromosomes in genetic algorithm.

Fig. 7.2 Sample particle

1	0	1	0	1

Fig. 7.3 QuickReduct step in PSO-QR

$$\forall x \in (C\text{-}R)$$
$$\gamma_{R\cup\{x\}}(D) = \frac{|POS_{R\cup\{x\}}(D)|}{|U|}$$

7.3.4 Incremental Feature Selection Algorithm (IFSA)

Incremental feature selection algorithm considers the dynamic dataset where attributes are added dynamically with passage of time. One of the positive aspects of the algorithm is that it provides explicit optimization step to remove the redundant attributes. The task is completed through measuring the significance of the attribute, i.e., dependency is calculated before and after removing attribute; if removal leads to decrease in dependency, it means attribute is not redundant and should not be removed. Algorithm, however, suffers from the following limitations:

- Significance is calculated using the expression: $sig_2(a,B,D) = \gamma B\cup\{a\}(D) - \gamma B$ (D), which means that we will have to calculate dependency twice, firstly with attribute "a" and secondly by excluding the attribute. This results serious performance degradation in case of large number of attributes.
- Algorithm converges to non-incremental feature selection algorithms in case of lesser number of features already known or more number of features added dynamically.

7.3.5 AFSA

Authors in [7] have presented a hybridized approach for feature selection using rough set theory and fish swarm logic. Algorithm uses swarm logic to iteratively optimize the solution to feature selection. Apart from the positive aspects of swarm logic, AFSA suffers the following limitations [8]:

- Higher time complexity.
- Lower convergence speed.
- Lack of balance between global search and local search.
- Not to use the experiences of group members for the next moves.
- Furthermore, there are no guidelines for input parameters. Efforts are made from time to time to improve these limitations.

The Table 7.3 presents some variations on basic fish swarm:

Table 7.3 PSO-based algorithms

PSO algorithm	Description
Feature selection and parameter optimization of support vector machines based on modified artificial fish swarm algorithms [9]	Research proposes a modified AFSA (MAFSA) to improve feature selection and parameter optimization for support vector machine classifiers
Feature optimization based on artificial fish swarm algorithm in intrusion detections [10]	Research proposes a method of optimization and simplification to network feature using artificial fish swarm algorithm in intrusion detection
Dataset reduction using improved fish swarm algorithm [11]	Research proposes a new intelligent swarm modelling approach that consists primarily of searching, swarming and following behaviours
Evolving neural network classifiers and feature subset using artificial fish swarm [12]	Research presents the use of AFSA as a new tool which sets up a neural network (NN), adjusts its parameters and performs feature reduction, all simultaneously
Feature selection for support vector machines base on modified artificial fish swarm algorithm [13]	Research proposes a modified version of artificial fish swarm algorithm to select the optimal feature subset to improve the classification accuracy for support vector machines

7.3.6 *Feature Selection Using Exhaustive and Heuristic Approach*

Authors in [14] propose a new feature selection approach taking advantage of both heuristic and exhaustive search algorithms. The approach in this way can avoid the limitations of both categories of search. Exhaustive search is impossible and takes too much time and thus cannot be used for large-scale datasets; heuristic search on the other hand does not provide the optimal results. In [14] authors first performed the heuristic approach to perform feature selection. After this they performed exhaustive search to remove the redundant features, thus ensuring the optimal feature subset.

The results were impressive as compared to bare use of exhaustive search; however, it suffers the following limitations:

- Algorithm experiences too much performance degradation as compared to other feature selection algorithms employing only one strategy.
- The limitations of the heuristic search algorithms like no explicit guidelines for input, no proper stopping criteria, etc., could not be removed.

7.3.7 *Feature Selection Using Random Feature Vectors*

In [15], authors have presented a feature selection approach based on hit and trial method. Research is based on randomly generating feature subsets until we get the one having maximum dependency, i.e. equal to that of the entire conditional attribute set. It then performs optimization step to remove the redundant features.

The proposed algorithm is better than [14] as it calculates feature vectors without using complex operators like GA, PSO, AFS, etc. However, algorithm suffers the following limitations:

- No explicit stopping criteria in case if hit and trial does not generate any ideal solution.
- Optimizing the resultant feature vector requires calculating dependency twice, once by including a feature and then by excluding it which may result in performance bottleneck.

7.4 Summary

In this chapter we have performed critical analysis of few of the most commonly used algorithms. Intention was to elaborate the strengths and weaknesses of each approach. For this purpose, firstly experimentation was conducted a benchmark dataset, and results were discussed. Then few of the algorithms were discussed in depth along with their strengths and weakness. This would give the research community a direction to further investigate these algorithms and overcome their limitations.

Bibliography

1. Raza MS, Qamar U (2016) An incremental dependency calculation technique for feature selection using rough sets. Inf Sci 343:41–65
2. Lichman M (2013) UCI machine learning repository [http://archive.Ics.Uci.Edu/ml]. University of California, School of Information and Computer Science, Irvine. Access date 30 Mar 2017
3. Jensen R, Shen Q (2008) Computational intelligence and feature selection: rough and fuzzy approaches. Wiley, Hoboken
4. Zuhtuogullari K, Allahvardi N, Arikan N (2013) Genetic algorithm and rough sets based hybrid approach for reduction of the input attributes in medical systems. Int J Innov Comput Inf Control 9:3015–3037
5. Inbarani HH, Azar AT, Jothi G (2014) Supervised hybrid feature selection based on PSO and rough sets for medical diagnosis. Comput Methods Prog Biomed 113(1):175–185
6. Bai Q (2010) Analysis of particle swarm optimization algorithm. Comput Inf Sci 3(1):180
7. Chen Y, Zhu Q, Xu H (2015) Finding rough set reducts with fish swarm algorithm. Knowl-Based Syst 81:22–29

8. Neshat M et al (2014) Artificial fish swarm algorithm: a survey of the state-of-the-art, hybridization, combinatorial and indicative applications. Artif Intell Rev 42(4):965–997
9. Lin K-C, Chen S-Y, Hung JC (2015) Feature selection and parameter optimization of support vector machines based on modified artificial fish swarm algorithms. Math Probl Eng 2015
10. Liu T et al (2009) Feature optimization based on artificial fish-swarm algorithm in intrusion detections. In: International Conference on Networks Security, Wireless Communications and Trusted Computing, 2009. NSWCTC'09, vol 1. IEEE
11. Manjupriankal M et al (2016) Dataset reduction using improved fish swarm algorithm. Int J Eng Sci Comput 6.4:3997–4000
12. Zhang M et al (2006) Evolving neural network classifiers and feature subset using artificial fish swarm. In: Mechatronics and automation, Proceedings of the 2006 I.E. International Conference on. IEEE
13. Lin K-C, Chen S-Y, Hung JC (2015) Feature selection for support vector machines base on modified artificial fish swarm algorithm. Ubiquitous Computing Application and Wireless Sensor. Springer, Dordrecht, pp 297–304.
14. Raza MS, Qamar U (2016) A hybrid feature selection approach based on heuristic and exhaustive algorithms using Rough set theory. In: Proceedings of the international conference on internet of things and cloud computing. ACM
15. Raza MS, Qamar U (2016) A rough set based feature selection approach using random feature vectors. In: Frontiers of Information Technology (FIT), 2016 international conference on. IEEE

Chapter 8
RST Source Code

In this chapter we will provide some implementation of some basic functions of rough set theory. Implementation of RST functions can be found in other libraries as well. The major aspect here is that the source code is also provided with each and every line explained. The explanation in this way will help research community to not only easily use the code but also they can modify as per their own research requirements. We have used Microsoft Excel VBA to implement the function. The reason behind is that VBA provides easy implementation and almost any of the dataset can easily be loaded in to the Excel. We will not only provide implementation of some of the basic RST concepts but also complete implementation and explanation of the source code of some of the most common algorithms like PSO, GA, QuickReduct, etc.

8.1 A Simple Tutorial

Before going into the details of source code, we will first explain some basic statements of Excel VBA that are most commonly used. Here we will provide very basic introduction about the syntax. For more details we will recommend you to take some good tutorial.

The online version of this chapter (doi:10.1007/978-981-10-4965-1_8) contains supplementary material, which is available to authorized users.

M.S. Raza, U. Qamar, *Understanding and Using Rough Set Based Feature Selection: Concepts, Techniques and Applications*,
DOI 10.1007/978-981-10-4965-1_8

8.1.1 Variable Declaration

Variables are declared with "Dim" statement. For example, to declare a variable by the name "Count" of type integer, we will use "Dim" statement as follows:

Dim count As Integer

Here "Dim" is the keyword to declare the variable, "Count" is the variable name, and "Integer" is the data type of "Count".

8.1.2 Array Declaration

Just like variable declaration, arrays are declared with "Dim" as keyword. For example, to declare a one-dimensional array by the name "List", we will use "Dim" statement as follows:

Dim List (2) As Integer

Here "Matrix" is name of array, and "2" is the upper bound of array bound of array, i.e. the index of last element. It should be noted that in Excel VBA arrays are zero-indexed, i.e. the first element has index "0", so above-defined array will have three elements. Same is the case with a two-dimensional array. Arrays in Excel VBA are dynamic, i.e. we can change dimensions at runtime; however, for that you have to define empty array (i.e. array without specifying array size) as follows:

Dim List() As Integer
ReDim List(3)
List(0) = 2

However, note that every time you redefine array, previous data will be lost.

For two-dimensional array, the same syntax will be followed but specifying two indexes as follows:

Dim Matrix(3,4) As Integer
Matrix(2,2) = 3

The above two lines define a two-dimensional array and initialized the element in third row and third column (remember that arrays are zero-indexed)

8.1.3 Comments

Comments are important part of any programming language. In Excel VBA comments are started by comma (') symbol. For example, the following line will be commented:

This is a comment and comments are turned green by default.

8.1.4 *If-Else Statement*

If-Else statement is a conditional statement used to implement branching. Its syntax is as follows:

```
If Count = 0 Then
'Statements here
Else
'Statements here
End If
```

After the keyword "If", there is expression to be evaluated, and "Then" is the keyword used. If the conditional expression solves to "True", statements till "Else" keyword will be executed otherwise statements after "Else" will be executed. Finally "End If" marks the end of "If" condition.

8.1.5 *Loops*

The most common loops used are "for-loop" and "while-loop". Below is the syntax of "for-loop":

```
For Index = 1 To 10
count = count +1
Next
```

Here "For" is the keyword, "Index" is the counter variable that starts with value "1", and loop will keep on iterating until the value of "Index" remains less than or equal to "10". Greater than "10" will cause the loop to terminate. "Next" marks end of the loop body.

For loop works as counter, while loop on the other hand keeps on iterating until a specific condition remains "True". The following is the syntax of while loop:

```
While (i < 10)
i = i + 1
Wend
```

"While" is the keyword after which we have the expression to be evaluated. Set of statements till the keyword "Wend" is called body and is executed until conditional expression remains "True".

8.1.6 *Functions*

Functions are a reusable code just like any other programming language. In Excel VBA function definition has the following syntax:

```
Function Sum(ByVal x As Integer, ByVal y As Integer) As Integer
Dim Answer As Integer
Answer = x + y
Sum = Answer
End Function
```

"Function" is the keyword, then there is name of the function, and parameter list is specified in parentheses. Here "ByVal" means "By Value"; to receive a parameter "By Reference", we will use the keyword "ByRef" while it will let us modify the original variable in case if any change is made in the function. After parentheses we specify the return type of the function. In the above example, the "Sum" function will return an integer. Then we have body of the function. To return a value from the function, it is assigned to the name of the function, e.g. here "Sum = Answer" means that the value of the variable "Answer" will be returned by function "Sum".

8.1.7 LBound and UBound Functions

LBound and UBound functions return the lower bound and upper bound of array. For example, for the array:

Dim List (3) as Integer
LBound (List) will return "0" and UBound (List) will return "3".

8.2 How to Import the Source Code

You are given the ".bas" files that contain MS Excel VBA code. In order to use and modify the code, ".bas" files need to be imported in an Excel file. Here we will explain how to import ".bas" files in Excel.

To use the source code, you need any of the MS Excel 2013 or later version. However, note that the document should enable macros in order to run the code. To enable a macro, perform the following steps.

1. Click File > Options (Fig. 8.1).
2. Click "Trust Center" from "Excel Options" dialogue (Fig. 8.2).
3. Click "Trust Center Setting" button on the same dialogue box (Fig. 8.3).

Alternatively you can save document as "Macro enabled" from file "Save" dialogue as follows (Fig. 8.4):

To write/update source code, click the Visual Basic button on Developer tab in ribbon as follows (Fig. 8.5):

Note: If "Developer" tab is not visible, then you can select it from "Excel Options" dialogue box from "Customize Ribbon" tab as follows (Fig. 8.6):

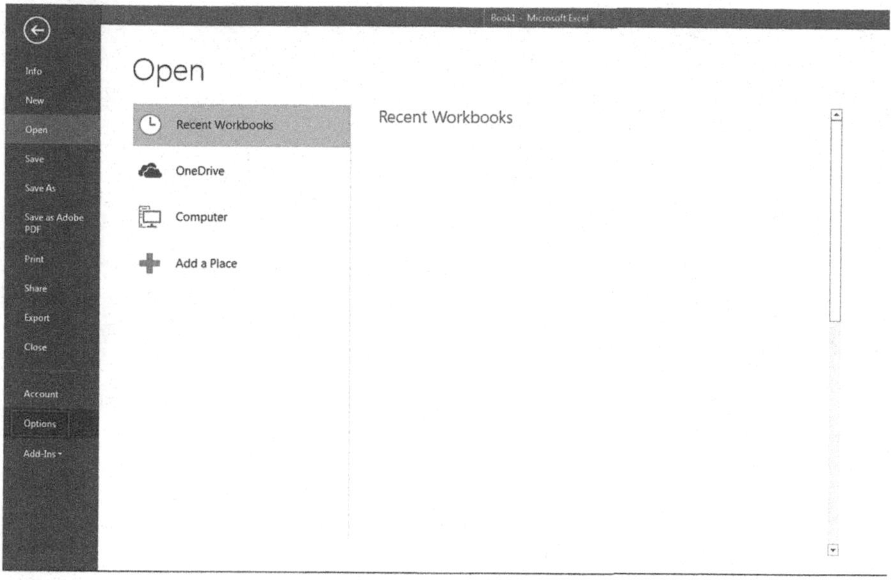

Fig. 8.1 File menu

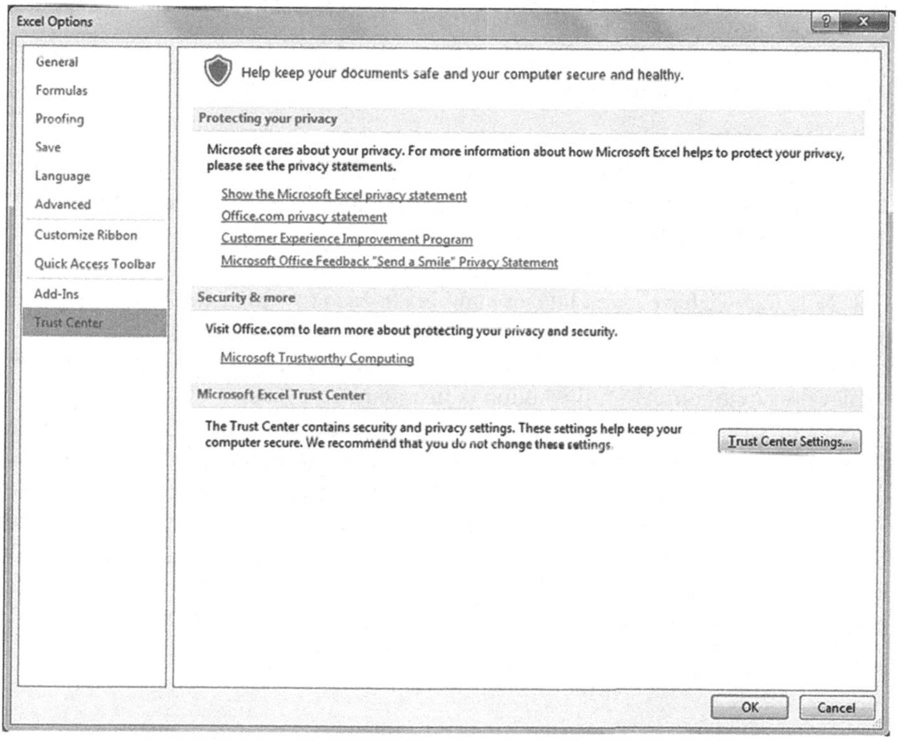

Fig. 8.2 Excel option dialogue box

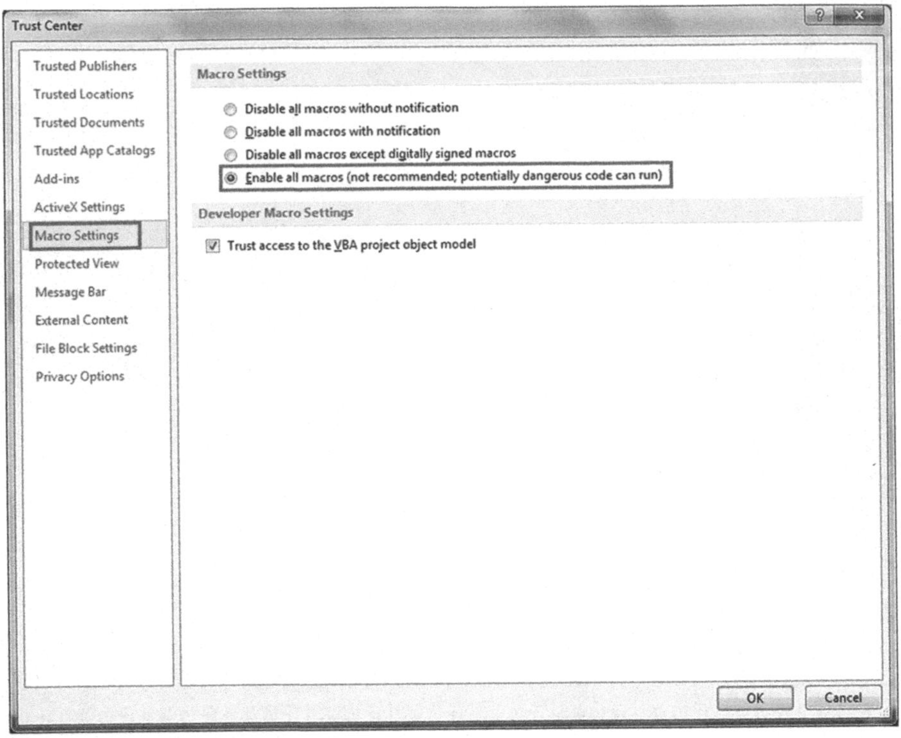

Fig. 8.3 Trust center dialogue

Before going into details of the source code, first of all let us explain the MS Excel files that contain source code. Here we explain structure of file using "Musk" dataset from (Fig. 8.7).

Actual dataset starts from "C3" cell. Column "B" specifies ObjectID of each row. Note that we have given integer numbers to each row to make the code simple. The last column, i.e. "FM", contains decision class (in case of Musk dataset). You can insert the command button ("FindDep" as shown in figure) from the "Insert" tab on developer ribbon. After, the button is inserted.

Now all you need is to import the given source code. For this purpose open "Visual Basic" tab from developer ribbon. Right click on the project name and click the "Import" submenu. From the open dialogue box, give the name of the ".bas" file and click "Open". The source code will be inserted. Now right click on the command button in the sheet and click "Assign Macro". Given the name "Main" as macro name and click OK. Your source code is ready to be executed. The table below shows the file name and functionality it implements (Table 8.1):

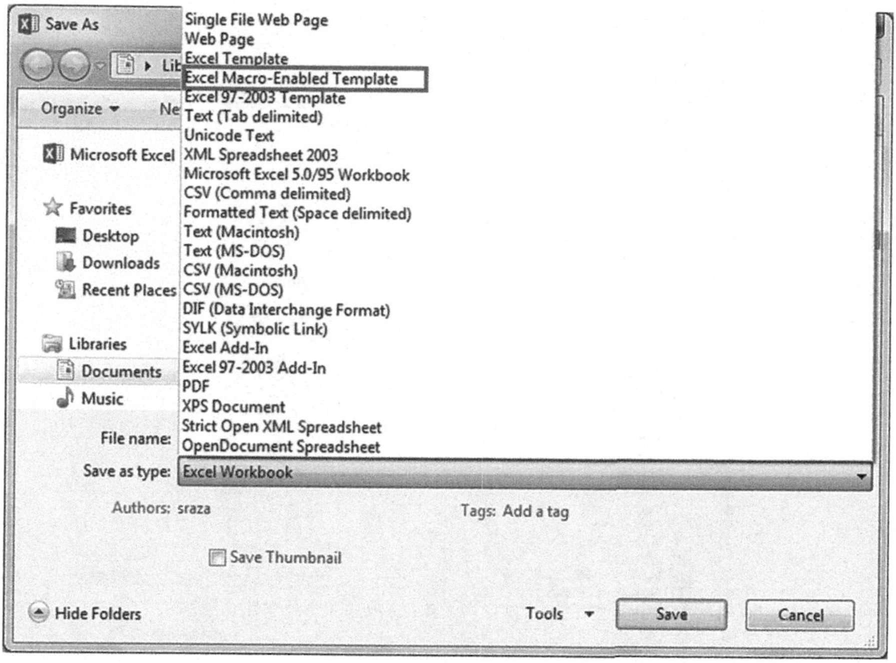

Fig. 8.4 Save dialogue

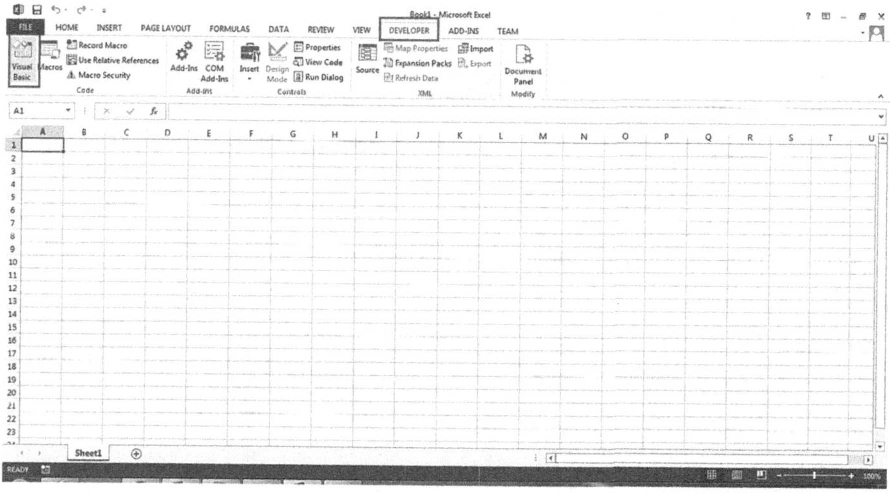

Fig. 8.5 Visual basic button in Developer tab

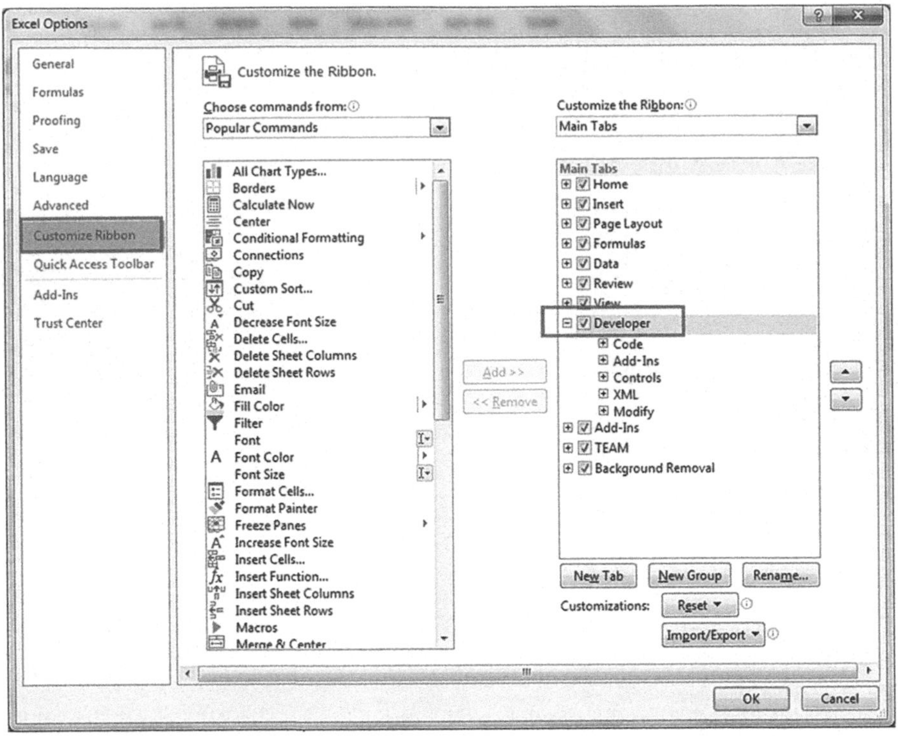

Fig. 8.6 Excel options dialogue

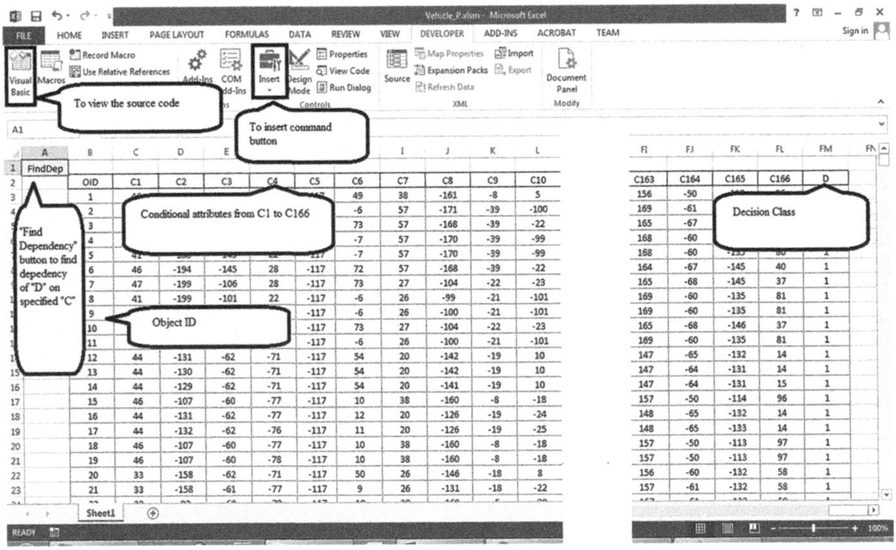

Fig. 8.7 How to store data in Excel

Table 8.1 Source code files and tasks they implement

File Name	Functionality Implemented
Musk_I_Dep.bas	Calculates dependency using incremental classes
Musk_P_Dep.bas	Calculates dependency using positive region
QuickReduct.bas	Implements QuickReduct algorithm
SpecHeart_I_LA.bas	Implements the calculation of lower approximation using dependency classes
SpecHeart_I_UA.bas	Implements the calculation of upper approximation using dependency classes
SpecHeart_P_LA.bas	Implements the calculation of upper approximation using conventional RST-based method
SpecHeart_P_UA.bas	Implements the calculation of lower approximation using conventional RST-based method

8.3 Calculating Dependency Using Positive Region

The button at the left top of the sheet captioned "FindDep" executes the code to find dependency. In this case dependency of the decision class "D" is found on conditional attributes mentioned in source code.

8.3.1 Main Function

Clicking the "FindDep" button executes the "Main" method. The code of main method is given in listing below (Listing 8.1):

Listing 8.1: Main Method

```
Function Main()
Dim i, j As Integer
Rub = 6598
Cub = 168
ReDim data(1 To rub, 1 To cub)
For i = 1 To rub
    For j = 1 To cub
Data(i, j) = Cells(2 + i, 1 + j).Value
    Next j
Next i
ObjectID = LBound(data, 2)
DAtt = UBound(data, 2)
Dim Chrom() As Integer
ReDim Chrom(1 To 150)
```

(continued)

Listing 8.1 (continued)

```
Row15:  For i = 2 To 10
Row16:  Chrom(i - 1) = i
Row17:  Next
Row18:  ReDim TotalCClasses(1 To RUB, 1 To RUB +1)
Row19:  dp = CalculateDRR(chrom)
Row20:  End Function
```

Now we will explain this code row by row. Row1 declares the main method. Then we have declared two variables i and j to use as index of the loops. RUB and CUB stand for "Row Upper Bound" and "Column Upper Bound". These two variables represent the maximum number of rows and columns in dataset. Note that there are 6598 rows in musk and there are 166 conditional attributes; however, we will have to specify two extra attributes, one for the decision class and one for the ObjectID. This will let us easily store data in our local array and refer to ObjectID just by specifying the first column of the array. Row5 declared an array called "Data" equal to total dataset size including decision class and ObjectID. As the interaction with Excel sheet is computationally too expensive at runtime, we will load the entire dataset in local array. Redim statement at Row5 specifies the size of the array. Array is initialized with the dataset from Row6 to Row10.

Row11 and Row12 initialize ObjectID and DAtt (Decision Attribute), respectively. ObjectID is initialized with index of the first column of "Data" array while DAtt is initialized with last index. Row14 defines an array named "Chrom" which actually contains the indexes of attributes on which dependency of decision class "D" is determined. Row15 to Row17 initializes chrom array with the attributes on which dependency of "D" is to be calculated. Note that Chrom will logically start from column "C" in sheet, i.e. the first conditional attribute "C1". So far this means that you have assigned consecutive attributes to find dependency; however, you can assign selective attributes, but in that case you will have to initialize the array manually without loop.

Row18 defines an array called "TotalCClasses" which basically stores the equivalence class structure using the conditional attributes. Note that "TotalCClasses" is defined globally and here it is assigned space dynamically. Finally Row19 calls the function "CalculateDRR(Chrom)" that calculates the dependency. Now we will explain "CalculateDRR" method.

8.3.2 Calculate DRR Function

CalculateDRR function actually calculates dependency of decision class on conditional attributes. Listing 8.2 shows the source code of this function; first we will explain working of this function and then all the functions called by it in sequence.

Listing 8.2: CalculateDRR Method

```
Row1:  Function CalculateDRR(ByRef chrom() As Integer) As Double
Row2:  Call SetDClasses
Row3:  Call ClrTCC
Row4:  Dim R As Integer
Row5:  Dim nr As Integer
Row6:  For R = 1 To RUB
Row7:  If (AlreadyExists(TotalCClasses, TCCRCounter, R) <> True)
       Then
Row8:  InsertObject Data(R, ObjectID), TotalCClasses, TCCRCounter
Row9:  For nr = R + 1 To RUB
Row10: If (MatchCClasses(R, nr, chrom, TCCRCounter) = True) Then
Row11: InsertObject Data(nr, ObjectID), TotalCClasses, TCCRCounter
Row12: End If
Row13: Next
Row14: TCCRCounter = TCCRCounter +1
Row15: End If
Row16: Next
Row17: Dim dp As Integer
Row18: Dim ccrx As Integer
Row19: Dim ddrx As Integer
Row20: dp = 0
Row21: For ccrx = 1 To TCCRCounter
Row22:    For ddrx = 1 To TDCRCounter
Row23:        dp = dp + FindDep(TotalCClasses, TotalDClasses, ccrx,
              ddrx)
Row24:    Next
Row25: Next
Row26: CalculateDRR = dp/RUB
Row27: End Function
```

Function accepts the reference of chrom array (or any array containing indexes of conditional attributes) and returns the dependency as a double value. The function first calls the "SetDClasses" function which constructs equivalence class structure using decision attribute. Please refer to the description of SetDClasses method in Sect. 8.3.3. Then it calls the function "ClrTCC" which initializes the array used for calculating the equivalence class structure developed by using conditional attributes. Please refer to the description of ClrTCC method in Sect. 8.3.5.

Then two local variables are declared with the name "R" and "NR" used as index for current now and next row. From Row7 to Row16, the function constructs equivalence class structure using conditional attributes. First loop considers each record and checks whether it already exists in equivalence class structure or not. In case it does not exist (i.e. this object is the one with whom no other object has same attribute values), it is inserted in equivalence class structure. Second loop starts with the next object from current object (selected in first loop) and matches values of conditional attributes for all remaining objects. All the objects that match are inserted in equivalence class structure. Please refer to the definitions of functions AlreadyExsits, InsertObject and MatchCClasses in Sects. 8.3.6, 8.3.7 and 8.3.8, respectively.

Row21 to Row25 calculate cardinality of positive region, i.e. third step of dependency calculation. First loop iterates through each equivalence class in "TotalCClasses" and sends it to function "PosReg" along with each equivalence class stored in "TotalDClasses". PosReg function checks whether the equivalence class in "TotalCClasses" is a subset of "TotalDClasses" or not. Once both loops complete, we have cardinality of objects stored in variable "dp". Please refer to the definition of PosReg method in Sect. 8.3.9. The cardinality is then divided by the total number of records in dataset to calculate and return actual dependency.

8.3.3 *SetDClasses Method*

This function calculates the equivalence class structure using decision attribute. Note that in calculating dependency using conventional positive region-based dependency measure, this is the first step. Listing 8.3 shows source code of this function.

Listing 8.3: SetDClasses Method

```
Row1:    Function SetDClasses() As Integer
Row2:    Dim C As Integer
Row3:    Dim Row As Integer
Row4:    Dim Ind As Integer
Row5:    Const MaxR As Long = 2
Row6:    Dim X As Integer
Row7:    TDCRCounter = 2
Row8:    X = 5581
Row9:    ReDim TotalDClasses(1 To TDCRCounter, 1 To (X + 3))
Row10:   TotalDClasses(1, 1) = 0
Row11:   TotalDClasses(1, 2) = 5581
Row12:   TotalDClasses(1, 3) = 3
Row13:   TotalDClasses(2, 1) = 1
Row14:   TotalDClasses(2, 2) = 1017
Row15:   TotalDClasses(2, 3) = 3
```

(continued)

Listing 8.3 (continued)

```
Row16:     For Row = 1 To RUB 'construct decision classes
Row17:        Ind = FindIndex(Row, TotalDClasses, maxr)
Row18:        TotalDClasses(Ind, TotalDClasses(Ind, 3) + 1) = Data(Row,
              ObjectID)
Row19:        TotalDClasses(Ind, 3) = TotalDClasses(Ind, 3) + 1
Row20:     Next
Row21:   SetDClasses = 1
Row22:   End Function
```

Row1 to Row7 define the local variables used. "TDCRCounter" represents the total number of decision classes. In Musk dataset, there are two decision classes, so we assigned it with value "2". MaxR represents the total number of rows in "TotalDClasses" array. This array contains rows equal to the number of decision classes, i.e. one row for each decision class. In our case there are two decision classes, so there will be two rows. The structure of "TotalDClasses" array is as follows (Fig. 8.8):

First column specifies the decision class, second column specifies total number of objects in this equivalence class, and third column stores index of last object. The rest of the columns store objects in that decision class. Here in this case objects X1, X2 and X4 belong to decision class "1".

The number of rows in "TotalDClasses" array is equal to the total number of decision classes and the number of columns equal to the maximum number of objects in any decision class plus "3".

Next the loop runs to create the equivalence class structure. It first finds the row number of the TotalDClasses array where the decision class (of the current record) is stored. It then places current object in the column next to last object in same row. Finally, it updates the total number of objects for that decision class and index of the last object.

8.3.4 *FindIndex Function*

This function finds the row number of the decision class of current object in "TotalDClasses" array. Code of the function is given below in Listing 8.4.

Fig. 8.8 Equivalence class structure using decision attribute

Decision class	Instance Count	Instance Count	Instances		
1	3	3	X1	X3	X4
2	1	3	X2		
3	1	3	X5		

Listing 8.4: FindIndex Method

```
Row1:   Function FindIndex(R As Integer, ByRef TDC() As Integer,
        MaxR As Integer) As
Row2:   Integer
Row3:   Dim C As Integer
Row4:       For C = 1 To MaxR
Row5:          If Data(R, DAtt) = TDC(C, 1) Then
Row6:             Exit For
Row7:          End If
Row8:       Next
Row9:      FindIndex = C
Row10:  End Function
```

Function takes three arguments, i.e. row number of current object, reference of "TotalDClasses" array and maximum number of rows in it. It then navigates through the first column of each row (where the decision class is stored). If the decision class is matched with the decision class of current object (whose ID is represented by variable R), then this object is returned. Note that if decision class of current object is not matched with any in "TotalDClasses" then index of last row is returned where this object will be stored.

8.3.5 ClrTCC Function

Listing 8.5 shows the source code of ClrTCC method.

Listing 8.5: ClrTCC Method

```
Row1:  Function ClrTCC() As Integer
Row2:  Dim i As Integer
Row3:  TCCRCounter = 0
Row4:  For i = 1 To RUB
```

(continued)

Listing 8.5 (continued)

```
Row5:   TotalCClasses(i, 1) = 0
Row6:   Next
Row7:   ClrTCC = 1
Row8:   End Function
```

Row1 defines the function. Variable "i" will be used as an index variable in the loop. First, let's explain the structure of how "TotalCClasses" stores equivalence class structure as shown in Fig. 8.9. Each row stores an equivalence class, e.g. objects X1 and X3 belong to the same equivalence class, whereas objects X2, X4 and X5 represent another equivalence class. The first column shows the total number of objects in an equivalence class, e.g. in the first row there are two objects in equivalence class, whereas in the second row there are three objects.

From Listing 7.1, note that the total number of rows in "TotalCClasses" is equal to the total number of rows in the dataset. This is done for the worst (very rare) cases where each object has its own decision class. Similarly, the number of columns is equal to the number of rows plus one (1); this is again for worst case where all objects belong to the same class. The extra column stores the total number of objects in each equivalence class.

Now the function "ClrTCC()" clears the first column of the "TotalCClasses" array. It then returns the value "1" to indicate that function has been successfully executed.

8.3.6 *AlreadyExists Method*

This method checks if an object already exists in equivalence classes. The following is the listing of this function (Listing 8.6).

Fig. 8.9 Equivalence class structure using conditional attributes

2	X1	X3		
3	X2	X4	X5	

Listing 8.6: AlreadyExists Method

```
Function AlreadyExists(ByRef TCC() As Integer, TCCRC As
Integer, R As Integer)
As Boolean
Dim i As Integer
Dim j As Integer
  Dim Exists As Boolean
  Exists = False
  If TCCRC = 0 Then
  Exists = False
  End If
 For i = 1 To TCCRC
   For j = 2 To TCC(i, 1) + 1
      If (TCC(i, j) = Data(R, ObjectID)) Then
      Exists = True
      Exit For
   End If
 Next
 If (Exists = True) Then
 Exit For
 End If
 Next
 AlreadyExists = Exists
 End Function
```

The function takes three arguments, the reference of "TotalCClasses" array, total number of rows in this array and row number (stored in "R") that is to be checked in equivalence classes. The function first checks if "TotalCClasses" array is empty in which case function returns "False". Then function runs two nested loops to check the objects column wise in each row. Note that the second loop starts with j = 2 because in the equivalence class array, the first column stores the total number of objects in the decision class and actual objects in an equivalence class start from the second column. If the object is found at any location, the function returns "True", or else "False".

8.3.7 InsertObject Method

This method inserts an object in equivalence class structure, i.e. "TotalCClasses". This function is called when object does not exist in array, so it inserts object in the next row and initializes the first column in this row equal to one (1). The following is the listing of this function (Listing 8.7).

Listing 8.7: InsertObject Method

```
Function InsertObject(O As Integer, ByRef TCC() As Integer,
TCCRC As Integer)
TCC(TCCRC +1, 1) = TCC(TCCRC +1, 1) + 1
TCC(TCCRC +1, TCC(TCCRC +1, 1) + 1) = O
End Function
```

Function takes three arguments, ObjectID of the object to be inserted, "TotalCClasses" array and total number of rows in the array so far having objects.

8.3.8 *MatchCClasses Function*

This function matches two objects according to value of their attributes. It takes three arguments. The following is the listing of this function (Listing 8.8).

Listing 8.8: MatchCClasses Method

```
Function MatchCClasses(R As Integer, NR As Integer, ByRef
chrom() As Integer,
TCCRC As Integer) As Boolean
Dim j As Integer
Dim Ci As Integer
Dim ChromSize As Integer
Dim ChromMatched As Boolean
ChromMatched = True
For Ci = LBound(chrom) To UBound(chrom)
 If (Data(R, chrom(Ci)) <> Data(NR, chrom(Ci))) Then
     ChromMatched = False
Exit For
End If
Next
MatchCClasses = ChromMatched
End Function
```

Function takes three arguments, ObjectIDs of the objects to be compared and the reference of "Chrom" array which contains attributes for which objects need to be compared. It runs the loop that matches the objects against the values of the attributes stored in Chrom array. If both objects have the same values against attributes mentioned in chrom, then it returns "True", or else "False".

8.3.9 PosReg Function

This function calculates cardinality of the objects belonging to the positive region. The listing of function is given below (Listing 8.9).

Listing 8.9: PosReg Method

```
Row1:   Function PosReg(ByRef TCC() As Integer, ByRef TDC() As
        Integer, cr As Integer, dr
Row2:   As Integer) As Integer
Row3:   Dim X, cnt, dpc, y As Integer
Row4:   dpc = 0
Row5:   cnt = TCC(cr, 1)
Row6:   If (TCC(cr, 1) < = TDC(dr, 2)) Then
Row7:   For X = 2 To TCC(cr, 1) + 1
Row8:   For y = 4 To TDC(dr, 3)
Row9:   If (TCC(cr, X) = TDC(dr, y)) Then
Row10:      dpc = dpc + 1
Row11:  End If
Row12:  Next
Row13:  Next
Row14:  End If
Row15:  If cnt = dpc Then
Row16:  PosReg = dpc
Row17:  Else
Row18:  PosReg = 0
Row19:  End If
Row20:  End Function
```

This function takes four arguments, i.e. the reference of "TotalCClasses" array, "TotalDClasses" array, current row in "TotalClasses" and current row in "TotalDClasses". The function then calculates the cardinality of objects in equivalence class stored in "TotalCClasses" (at row number "cr") which are subset of equivalence class sored in "TotalDClasses" (at row number "dr"). Note that the first loop at Row7 starts with X = 2 because in "TotalCClasses" objects in each equivalence class start from index 2. Similarly in "TotalDClasses", indexes of objects start from index 4.

8.4 Calculating Dependency Using Incremental Dependency Classes

Now we will explain how dependency can be calculated using IDCs. The data will be stored in Excel file the same as we stored in the case of positive region-based dependency measure.

8.4.1 Main Function

Clicking the "FindDep" button will call "Main" method. Loading the data in local array "Data" and other variables is the same as discussed before. The following is the listing of the Main function (Listing 8.10).

Listing 8.10: Main Method

```
Row1:  Function Main()
Row2:  Dim i, j As Integer
Row3:  Rub = 6598
Row4:  Cub = 168
Row5:  ReDim data(1 To rub, 1 To cub)
Row6:  For i = 1 To rub
Row7:      For j = 1 To cub
Row8:  Data(i, j) = Cells(2 + i, 1 + j).Value
Row9:      Next j
Row10:  Next i
Row11:  ObjectID = LBound(data, 2)
Row12:  DAtt = UBound(data, 2)
Row13:  Dim Chrom() As Integer
Row14:  ReDim Chrom(1 To 150)
Row15:  For i = 2 To 10
Row16:  Chrom(i - 1) = i
Row17:  Next
Row18:  dp = CalculateDID(chrom)
Row19:  End Function
```

All of the function is the same as discussed before; however, at Row18, the function CalculateDID will calculate dependency using IDCs.

8.4.2 CalculateDID Function

CalculateDID calculates dependency using dependency classes. Listing 8.11 shows the source code of the function.

Listing 8.11: CalculateDID Method

```
Function calculateDID(ByRef chrom() As Integer) As Double
Dim DF As Integer
Dim UC As Integer
Dim ChromSize As Integer
Dim FoundInGrid As Boolean
Dim i As Integer
Dim GRC As Integer
Dim ChromMatched As Boolean
Dim DClassMatched As Boolean
Dim ChromMatchedAt As Integer
Dim DClassMatchedAt As Integer
GridRCounter = 0
FoundInGrid = False
ChromMatched = False
DClassMatched = False
ChromSize = UBound(chrom) - LBound(chrom) + 1
DECISIONCLASS = UBound(chrom) + 1
INSTANCECOUNT = DECISIONCLASS +1
AStatus = INSTANCECOUNT +1
ReDim Grid(1 To ChromSize +3, 1 To RUB)
If (GridRCounter = 0) Then
    GridRCounter = Insert(GridRCounter, chrom, 1)
    DF = 1
    UC = 1
End If
For i = 2 To RUB
    ChromMatchedAt = MatchChrom(i, chrom, ChromMatched,
    GridRCounter)
    If (ChromMatched = True) Then
        If (Grid(AStatus, ChromMatchedAt) <> 1) Then
            DClassMatched = MatchDClass(i, ChromMatchedAt)
            If (DClassMatched = True) Then
                DF = DF + 1
                Grid(INSTANCECOUNT, ChromMatchedAt) =
Grid(INSTANCECOUNT, ChromMatchedAt) + 1
            Else
                DF = DF - Grid(INSTANCECOUNT, ChromMatchedAt)
                Grid(AStatus, ChromMatchedAt) = 1
            End If
        End If
    Else
```

(continued)

Listing 8.11 (continued)

```
Row40:        GridRCounter = Insert(GridRCounter, chrom, i)
Row45:        DF = DF + 1
Row46:      End If
Row47:    UC = UC + 1
Row48:   Next
Row49:   calculateDID = DF/UC
         End Function
```

Function takes reference of integer array "Chrom" as input. As discussed earlier "Chrom" contains indexes of the conditional attributes on which the dependency of "D" is to be determined. "CalculateDID" uses "Grid" as intermediate data structure to calculate dependency. "Grid" is just a two-dimensional array with a number of rows equal to the number of records in dataset and a number of columns equal to the total number of conditional attributes on which dependency is to be determined plus "3", e.g. if dependency is to be determined on attributes "A1, A3 and A3", the number of columns should be equal to "6". The last column specifies status of the attribute set that either these values of attributes have already been considered or not. Second last column specifies the total number of occurrences of the current value set in dataset so that if the same value of these attributes leads to a different decision class later, we may subtract this number from current dependency value. Third last column specifies the decision class and the first n-3 columns specify the values under current attribute set. This matrix is filled for all the instances in dataset. So for the dataset given in Table 8.2:

Contents of the Grid will be as follows (for first three objects) (Fig. 8.10):

In Row22, we insert the first record in the Grid. Please refer to the definition of "Insert" method (Sect. 8.4.3) about how to insert a record. "DF" and "UC" store "Dependency Factor" and "Universe Count". Dependency factor stores the total number of objects corresponding to positive region in conventional method. As so far we have only inserted the first record, so both DF and UC will be equal to "1". Next method starts a loop from i = 2 (because we have already inserted the first record).

In Row27, MatchChrom function matches that if any object with the same attribute values as that of object "i" already exists in the Grid or not. In case if the object with the same value already exists in the Grid, we will check whether we already have considered it or not (i.e. either we have updated DF on the basis of this object or not); if this object has not been considered previously, we will match the decision class of the object with the one stored in the Grid. For this purpose "MatchDClass" function is used. If decision class is also matched, we increment the "DF" factor which means that this object belongs to a positive region. In Row33, we increment the instance count (second last) column. However, if the decision class does not match, then we will decrement the DF by the factor equal to the total number of previous occurrences of all objects with same attribute values.

Table 8.2 Sample decision system

U	State	Qualification	Job
x_1	S1	Doctorate	Yes
x_2	S1	Diploma	No
x_3	S2	Masters	No
x_4	S2	Masters	Yes
x_5	S3	Bachelors	No
x_6	S3	Bachelors	Yes
x_7	S3	Bachelors	No

Fig. 8.10 Grid for calculating IDC

S1	Doctorate	Yes	1	1
S1	Diploma	No	1	1
S2	Masters	No	1	1

We will also set last column (AStatus) to "1" to indicate that object has already been considered. However, if this object has already been considered, we will simply increment UC factor. In case if no object with the same attribute values already exists in Grid, we will insert this object using "Insert" function and will update "DF".

Finally in Row49 we calculate and return actual dependency value.

8.4.3 *Insert Method*

This method inserts an object in Grid. Listing 8.12 shows the source code of this method.

Listing 8.12: Insert Method

```
Function Insert(GRC As Integer, chrom() As Integer, drc As Integer) As Integer
Dim Ci As Integer
For Ci = 1 To UBound(chrom)
Grid(Ci, GRC + 1) = Data(drc, chrom(Ci))
Next
  Grid(Ci, GRC + 1) = Data(drc, DAtt)
Grid(Ci + 1, GRC + 1) = 1 'instance count
Grid(Ci + 2, GRC + 1) = 0 'status
  Insert = GRC + 1
End Function
```

The function receives three arguments, i.e. grid record count (GRC) is the total number of records already inserted in Grid; note that Insert method is called when a record is inserted in the Grid for the first time, so we need to insert it in the next row after the last inserted row in the Grid. It then iterates through all the attribute indexes stored in Chrom array and stores them in first n-3 columns of the Grid. In third last column, the decision class is stored; in the second last column, instance count is set to "1" as the record with such attribute values appears in the Grid for the first time and finally status set to "0" means that this object has not been considered before.

8.4.4 MatchChrom Method

MatchChrom method compares the attribute values of current object with all those in the Grid. The following is the listing of this method (Listing 8.13).

Listing 8.13: MatchChrom Method

```
Function MatchChrom(i As Integer, ByRef chrom() As Integer,
ByRef ChromMatched As Boolean, GRC As Integer) As Integer
Dim j As Integer
Dim Ci As Integer
   For j = 1 To GRC
      ChromMatched = True
     For Ci = 1 To UBound(chrom)
If (Data(i, chrom(Ci)) <> Grid(Ci, j)) Then
ChromMatched = False
Exit For
End If
     Next
     If (ChromMatched = True) Then
Exit For
     End If
   Next
MatchChrom = j
End Function
```

It receives three arguments, i.e. grid record count (GRC), reference of the chrom array and data record count (drc) that stores the index of the object whose attribute values need to be compared. Function iterates through rows, starting from first row in Grid, it compares attribute values with that of current object, first the attribute values stored in first row are compared, then that in second row and so on. At any row number where attribute values are match, that row number is returned.

8.4.5 MatchDClass Method

In CaclculateDID method, MatchDClass is called if attribute values are compared and the row number where the attribute values are compared are returned. Now at that row in the Grid, MatchDClass compares the decision class stored with that of the current object. The following is the listing of this function (Listing 8.14).

Listing 8.14: MatchDClass Method

```
Row1:   Function MatchDClass(i As Integer, ChromMatchedAt As Integer)
        As Boolean
         Dim DCMatched As Boolean
Row2:   DCMatched = False
Row3:      If (Data(i, DAtt) = Grid(DECISIONCLASS, ChromMatchedAt))
           Then
Row4:         DCMatched = True
Row5:      End If
Row6:      MatchDClass = DCMatched
Row7:   End Function
```

It receives two arguments, i.e. the row number of the object in the dataset and the row number in the Grid where the chrom was matched. If the decision class matches with that of the object in dataset, then the function returns "True", or else "False".

Table 8.3 shows the important variables and their description.

8.5 Lower Approximation Using Conventional Method

The following function can be used to calculate lower approximation using conventional indiscernibility-based method.

8.5.1 Main Method

Listing 8.15 shows the source code of Main method.

Table 8.3 Variables and their description

Variable	Datatype	Purpose
Data()	Integer	Integer array that stores data. First column of the array stores ObjectID and last column stores decision class
ObjectID	Integer	Refers to the index of first column in "Data" array. Used to refer to a particular object
DAtt	Integer	Stores index of column containing decision class
RUB	Integer	Stores count of the total number of rows in dataset
CUB	Integer	Stores total number of columns in dataset
TotalDClasses()	Integer	Stores equivalence class structure using decision class
TotalCClasses()	Integer	Stores equivalence class structure using conditional attributes
TDCRCounter	Integer	Total number of equivalence classes in "TotalCClasses" array
TCCRCounter	Integer	Total number of equivalence classes in "TotalDClasses" array

Listing 8.15: Main Method

```
Function Main() As Integer
GridRCounter = 0
Dim i, j As Integer
RUB = 80
CUB = 46
ReDim data(1 To RUB, 1 To CUB)
For i = 1 To RUB
  For j = 1 To CUB
    data(i, j) = Cells(2 + i, 1 + j).Value
  Next j
Next i
OID = LBound(data, 2)
DAtt = UBound(data, 2)
Dim chrom() As Integer
ReDim chrom(1 To CUB - 2)
For i = 2 To CUB - 1
chrom(i - 1) = i
Next
CalculateLAObjects chrom, 1
Dim str As String
For i = 1 To UBound(LAObjects)
  If LAObjects(i) <> 0 Then
  str = str & ",X" & LAObjects(i) & ""
  End If
Next
```

(continued)

Listing 8.15 (continued)

```
Row27:   Cells(2, 1).Value = str
Row28:   Main = 1
Row29:   End Function
Row30:
```

Function loads the data in the same way as we have previously explained. At Row19 it calls "CalculateLAObjects" to calculate objects belonging to lower approximation. The function fills "LAObjects" array with all such objects. This array contains indexes (record number) of objects belonging to lower approximation. Row21 to Row26 iterates through each object and stores them in a string which is then shown in Cell "A2". This is just for the display purpose; however, for any further processing on objects belonging to lower approximation, "LAObjects" can be used.

8.5.2 *CalculateLAObjects Method*

This method calculates equivalence class structure using conditional attributes. The following is the listing of this method (Listing 8.16).

Listing 8.16: CalculateLAObjects Method

```
Row1:    Function CalculateLAObjects(ByRef chrom() As Integer, ByRef xc
         as Integer)
Row2:    Dim i As Integer
Row3:    Dim j As Integer
Row4:    Dim found As Integer
Row5:    SetDConcept xc
Row6:    Dim TotalCClasses() As Integer
Row7:    Dim TCCRCounter As Integer
Row8:    TCCRCounter = 0
Row9:    ReDim TotalCClasses(1 To RUB, 1 To RUB +1)
Row10:     Dim R As Integer
Row11:     Dim nr As Integer
Row12:     For R = 1 To RUB
Row13:     If (AlreadyExists(TotalCClasses, TCCRCounter, R, data) <>
           True) Then
Row14:     InsertObject data(R, OID), TotalCClasses, TCCRCounter
Row15:     For nr = R + 1 To RUB
Row16:     If (MatchCClasses(R, nr, chrom, TCCRCounter) = True) Then
Row17:     InsertObject data(nr, OID), TotalCClasses, TCCRCounter
```

(continued)

Listing 8.16 (continued)

```
Row18:    End If
Row19:    Next
Row20:        TCCRCounter = TCCRCounter +1
Row21:    End If
Row22:    Next
Row23:    Dim ccrx As Integer
Row24:    ReDim LAObjects(1 To RUB +1)
Row25:    LAObjects(1) = 0
Row26:    For ccrx = 1 To TCCRCounter
Row27:      FindLAO TotalCClasses, ccrx
Row28:    Next
Row29:    ccrx = 0
Row30:    End Function
```

It takes two arguments, the "Chrom" array, which contains indexes of the attributes under consideration and the "XConcept". "XConcept" specifies the concept (in simple words, the value of decision class) for which we need to determine lower approximation. First the function calls "SetDConcept" method and passes it the value of "XConcept" (Please refer to the definition of "SetDConcept" method in Sect. 8.5.4). It then creates an array "TotalCClasses" in the same way as discussed in previous sections. From Row12 to Row22, it constructs the equivalence class structure using conditional attributes mentioned in the "Chrom" array. The variable "TCCRcounter"stores the total number of equivalence classes in "TotalCClasses" array. It then calls the function FinLAO to which it actually performs the third step of finding lower approximations, i.e. all the objects in equivalence class structure (using conditional attributes) which are subset of equivalence class structure (using decision attribute) of XConcept. For example, if there are three equivalence classes in "TotalCClasses" stored in three rows, then the "FindLAO" will be called for each equivalence class. The following is the description of "FindLAO" method.

8.5.3 FindLAO Method

Find "LAO" method actually finds the objects belonging to lower approximation. The following is the listing of this method (Listing 8.17).

Listing 8.17: FindLAO Method

```
Row1:   Function FindLAO(ByRef TCC() As Integer, cr As Integer)
Row2:   Dim X, cnt, lac, y As Integer
Row3:   lac = 0
Row4:   cnt = TCC(cr, 1)
Row5:   If (TCC(cr, 1) < = XDConcept(1)) Then
Row6:       For X = 2 To TCC(cr, 1) + 1
Row7:          For y = 2 To XDConcept(1) + 1
Row8:             If (TCC(cr, X) = XDConcept(y)) Then
Row9:   lac = lac +1
Row10:               End If
Row11:     Next
Row12:  Next
Row13:  End If
Row14:  If cnt = lac Then
Row15:  For X = 2 To TCC(cr, 1) + 1
Row16:        LAObjects(1) = LAObjects(1) + 1
Row17:        LAObjects(LAObjects(1) + 1) = TCC(cr, X)
Row18:     Next
Row19:  End If
Row20:  End Function
```

It takes two arguments, i.e. the reference of "TotalCClasses" array and the row number, i.e. the row index of the equivalence class structure which needs to be determined either as subset of "XConcept" or not. In Row4, it first determines the total number of objects in current equivalence class structure, if the number of objects in current equivalence class structure are less than those in equivalence class structure using decision attribute, then it runs two nested for loops, first loop picks and object and iterates through equivalence structure of "XConcept", if at any location the object is found, we increment lower approximation count (lac) variable. Finally we check if this count is less than the total number of objects belonging to XConcept, or not. In case if the total number of objects in current equivalence class is less than those belonging to XConcept, it means that a set of objects in this equivalence class is subset of XConcept and thus belongs to lower approximation. Then we copy all these objects to "LAObjects" array.

8.5.4 SetDConcept Method

This method constructs equivalence class structure using XConcept. The following is the listing of this function (Listing 8.18).

Listing 8.18: SetDConcept Method

```
Function SetDConcept (ByRef xc as Integer) As Integer
Dim XConcept As Integer
Dim i As Integer
XConcept = xc
ReDim XDConcept(1 To RUB +1)
Dim CDCI As Integer
CDCI = 2
XDConcept(1) = 0
For i = 1 To RUB
    If (data(i, DAtt) = XConcept) Then
        XDConcept(1) = XDConcept(1) + 1
        XDConcept(CDCI) = data(i, OID)
        CDCI = CDCI +1
    End If
Next
SetDConcept = 1
End Function
```

Function takes single argument, i.e. the XConcept value for which we need to construct equivalence class structure. Function then declares an XDConccpts array which will store the objects belonging to XConcept. Note that it is a one-dimensional array with size equal to the total number of records in dataset plus 1 (in case if all objects in a dataset belong to the same class). The extra column (first one) stores the total number of objects in XDConcepts. This function traverses through all the records and stores index of those object (in XDConcepts array) for whom decision class values match XConcept. It also updates the first index of SDConcepts array to represent total number of objects in it.

8.6 Lower Approximation Using Redefined Preliminaries

Lower approximation is calculated by "CalculateLAI" method. This method calculates lower approximation using redefined preliminaries. The following is the listing of this function (Listing 8.19).

Listing 8.19: CalculateLAI Method

```
Function CalculateLAI(ByRef chrom() As Integer, ByVal xc As Integer) As Single
Dim DF, UC As Integer.
Dim i As Integer
Dim GRC As Integer
Dim ChromMatched As Boolean
Dim DClassMatched As Boolean
Dim ChromMatchedAt As Integer
Dim DClassMatchedAt As Integer
Dim NObject As Integer
ReDim Grid(1 To cub +3, 1 To rub)
GridRCounter = 0
ChromMatched = False
DClassMatched = False
ChromSize = UBound(chrom) - LBound(chrom) + 1
DECISIONCLASS = UBound(chrom) + 1
INSTANCECOUNT = DECISIONCLASS +1
AStatus = INSTANCECOUNT +1
XConcept = xc
TDO = 0
For i = 1 To rub
   If (data(i, DAtt) = XConcept) Then
      TDO = TDO + 1
   End If
Next
ReDim Grid(1 To ChromSize +3 + TDO, 1 To rub) 'column,row
If (GridRCounter = 0) Then
   GridRCounter = Insert(GridRCounter, chrom, 1) '1 represents first record
End If
For i = 2 To rub
      ChromMatchedAt = MatchChrom(i, chrom, ChromMatched, GridRCounter)
If (ChromMatched = True) Then
   Grid(INSTANCECOUNT, ChromMatchedAt) = Grid(INSTANCECOUNT, ChromMatchedAt) + 1,
If (Grid(AStatus, ChromMatchedAt) = 0) Then
         Grid((ChromSize +2 + Grid(INSTANCECOUNT,
ChromMatchedAt) +1), ChromMatchedAt) = i
   End If
   If (MatchDClass(i, ChromMatchedAt) = False) Then
```

(continued)

Listing 8.19 (continued)

```
Row38:              Grid(AStatus, ChromMatchedAt) = 1
Row39:        End If
Row40:    Else
Row41:              GridRCounter = Insert(GridRCounter, chrom, i)
Row42:    End If
Row43:    Next
Row44:    Dim j As Integer
Row45:    ReDim t(1 To rub +1)
Row46:    t(1) = 0
Row47:    For i = 1 To GridRCounter
Row48:        If ((Grid(AStatus, i) = 0) And ((Grid(DECISIONCLASS, i) =
              1))) Then
Row49:    For j = 1 To Grid(INSTANCECOUNT, i)
Row50:    t(1) = t(1) + 1
Row51:    t(t(1) + 1) = Grid((ChromSize +3 + j), i)
Row52:          Next
Row53:        End If
Row54:    Next
Row55:    CalculateUAI = 0
Row56:    End Function
```

Function takes input two parameters, i.e. Chrom array and XConcept value (in xc variable). In Row10, function declares a Grid array and sets the initial grid record count (GridRCounter) to zero. ChromSize represents the total number of attributes in Chrom array. DECISIONCLASS, INSTANCECOUNT and AStatus store the index of the column that stores decision class, index of column that stores total number of objects in current decision class and finally the status of the object (either it has been considered or not). Note that this time, the function offers a bit of optimization in a way that it first determines the total number of objects that belong to XConcept and declares Grid according to that.

In Row27 first record is inserted into the Grid. Values of all the attributes are inserted, instance count is set to "1", attribute status is set to zero, and index of the object (from dataset). Function then traverses all the records in dataset from record number "2" to the last record, and for each record it matches the value of the attributes in dataset with those stored in the Grid. If values of the attributes are matched then the value of decision class is also checked. If decision class also matches, then the objects are stored in the next column in the same row (in the Grid) and instance count is incremented. Otherwise if decision class is not matched, then "AStatus" attribute is set to "1" which means that all the objects with the same value of attributes will not be part of lower approximation. However, if object with same value of attributes are not matched with those stored in Grid, the object is inserted in the same way as first record and GridRCounter is incremented.

After traversing all the records in dataset, the Grid now contains values of conditional attributes (in row), the number of objects having these values, decision class and finally the indexes of all the objects that contain this decision class. Now from Row47 to Row54 we run two nested loops. First loop iterates through rows and inner loops iterates through columns in each row. In each row, first loop checks if the "AStatus" column is "0" (which means that objects in this row belong to lower approximation) and decision class is same as that mentioned in XConcept (which means that objects belong to lower approximation of same decision class. The objects are stored in a temporary array "t". You can store it in any global array or alternatively array can be passed to function as reference and this function will fill the array.

8.7 Upper Approximation Using Conventional Method

Upper approximation is calculated by "FindUAO" method. This method is called from "CalculateUAObjects" which is the same as "CalculateLAObjects"; however, in final steps instead of calculating objects in equivalence class structure (using conditional attributes) which are subset of XConcept set, we find those equivalence classes which have nonempty interaction with XConcept set. This is what is implemented in FindUAO method (Listing 8.20).

Listing 8.20: FindUAO Method

```
Function FindUAO(ByRef TCC() As Integer, cr As Integer)
Dim XAs Integer
Dim IsNonNegative As Boolean
IsNonNegative = False
Dim i As Integer
For i = 2 To TCC(cr, 1) + 1
    If (data(TCC(cr, i), DAtt) = XConcept) Then
        IsNonNegative = True
    End If
 Next
    If IsNonNegative = True Then
        For X = 2 To TCC(cr, 1) + 1
        UAObjects(1) = UAObjects(1) + 1
        UAObjects(UAObjects(1) + 1) = TCC(cr, X)
        Next
    End If
 End Function
```

It receives two arguments, i.e. reference of "TocalCClasses" array and row index of current equivalence class. From Row6 to Row10, it iterates through equivalence class and checks whether any of the objects in "TotalCClasses" (for current equivalence class) has the same decision class value as mentioned in the concept. If this is the case then all the objects belonging to this equivalence class will be part of upper approximation and will be assigned to "UAObjects" array.

8.8 Upper Approximation Using Redefined Preliminaries

CalculateUAI method calculates upper approximation using redefined preliminaries. The method is called from the Main method. The following is the listing of this method (Listing 8.21).

Listing 8.21: CalculateUAI Method

```
Row1:   Function CalculateUAI(ByRef chrom() As Integer, ByRef xc As
        Integer) As Single
Row2:   Dim DF, UC As Integer
Row3:   Dim i As Integer
Row4:   Dim GRC As Integer
Row5:   Dim ChromMatched As Boolean
Row6:   Dim DClassMatched As Boolean
Row7:   Dim ChromMatchedAt As Integer
Row8:   Dim DClassMatchedAt As Integer
Row9:   Dim NObject As Integer
Row10:  ReDim Grid(1 To cub +3, 1 To rub)
Row11:  GridRCounter = 0
Row12:  ChromMatched = False
Row13:  DClassMatched = False
Row14:  ChromSize = UBound(chrom) - LBound(chrom) + 1
Row15:  DECISIONCLASS = UBound(chrom) + 1
Row16:  INSTANCECOUNT = DECISIONCLASS +1
Row17:  AStatus = INSTANCECOUNT +1
Row18:  XConcept = xc
Row19:  TDO = 0
Row20:  For i = 1 To rub
Row21:     If (data(i, DAtt) = XConcept) Then
Row22:        TDO = TDO + 1
Row23:     End If
Row24:  Next
Row25:  ReDim Grid(1 To ChromSize +3 + TDO, 1 To rub) 'column,row
```

(continued)

Listing 8.21 (continued)

```
Row26:  If (GridRCounter = 0) Then
Row27:     GridRCounter = Insert(GridRCounter, chrom, 1)
Row28:  End If
Row29:  For i = 2 To rub
Row30:        ChromMatchedAt = MatchChrom(i, chrom, ChromMatched,
              GridRCounter)
              If (ChromMatched = True) Then
Row31:          Grid(INSTANCECOUNT, ChromMatchedAt) = Grid
                (INSTANCECOUNT,
Row32:  ChromMatchedAt) + 1
Row33:           Grid((ChromSize +2 + Grid(INSTANCECOUNT,
                 ChromMatchedAt) + 1), ChromMatchedAt) = i
Row34:           If (MatchDClass(i, ChromMatchedAt) = False) Then
Row35:              Grid(AStatus, ChromMatchedAt) = 1
Row36:           End If
Row37:        Else
Row38:              GridRCounter = Insert(GridRCounter, chrom, i)
Row39:        End If
Row40:  Next
Row41:  Dim j As Integer
Row42:  ReDim t(1 To rub +1)
Row43:  t(1) = 0
Row44:  For i = 1 To GridRCounter
Row45:  Dim isnonnegative As Boolean
Row46:  isnonnegative = False
Row47:  For j = 1 To Grid(INSTANCECOUNT, i)
Row48:  If (data(Grid((ChromSize +3 + j), i), DAtt) = XConcept) Then
Row49:  isnonnegative = True
Row50:  End If
Row51:  Next
Row52:  If isnonnegative = True Then
Row53:  For j = 1 To Grid(INSTANCECOUNT, i)
Row54:  t(1) = t(1) + 1
Row55:  t(t(1) + 1) = Grid((ChromSize +3 + j), i)
Row56:  Next
Row57:  End If
Row58:  Next
Row59:  CalculateUAI = 0
Row60:  End Function
```

Just like "CalculateLAI", "CalulateUAI" takes the same arguments, i.e. reference of the chrom array and XConcept. Function performs same steps and stores objects in the Grid with same structure. However, note that now we have to calculate upper approximation, so all the objects having same value of decision attributes and at least one of them leads to same decision class specified by XConcept will be part of upper approximation. So, note that we have skipped the condition to Check "AStatus" only decision class is checked. But there is a complexity; an object inserted later (in the Grid) with the same value of conditional attributes may lead to a different decision class (and update the decision class column as well), so we have used two nested loops: the first loop checks the decision class of each object (in Grid) to ensure that either of them leads to the same decision class (as mentioned in XConcept); if any of such objects is found, then all the objects in the same row will be part of upper approximation.

8.9 QuickReduct Algorithm

QuickReduct algorithm is one of the most famous feature selection algorithms, so we have provided the complete source code. Clicking the execute button will invoke the QuickReduct method. The following is the listing of this method (Listing 8.22).

Listing 8.22: QuickReduct Method

```
Function QuickReduct () As Integer
GridRCounter = 0
Dim i, j As Integer
RUB = 6598
CUB = 168
ReDim data(1 To RUB, 1 To CUB)
For i = 1 To RUB
  For j = 1 To CUB
     data(i, j) = Cells(2 + i, 1 + j).Value
  Next j
Next i
OID = LBound(data, 2)
DAtt = UBound(data, 2)
Dim chrom() As Integer
ReDim chrom(1 To CUB - 2)
For i = 2 To CUB - 1
chrom(i - 1) = i
Next
Dim R() As Integer
Dim T() As Integer
Dim X() As Integer
```

(continued)

Listing 8.22 (continued)

```
Row22:  Dim tmp(1 To 1) As Integer
Row23:  Dim dp, sc, dp1, dp2, dp3 As Integer
Row24:  ReDim TotalCClasses(1 To RUB, 1 To RUB +1)
Row25:  Call ClrTCC
Row26:  Call setdclasses
Row27:  dp = calculateDRR(chrom)
Row28:  i = 1
Row29:  Do
Row30:  Range("A2").Value = i
Row31:  Call Restore(T, R)
Row32:  Call C_R(X, chrom, R)
Row33:  For sc = 1 To UBound(X)
Row34:  tmp(1) = X(sc)
Row35:  Call ClrTCC
Row36:  dp1 = calculateDRR(CUD(R, tmp))
Row37:  If (dp1 > dp2) Then
Row38:  T = CUD(R, tmp)
Row39:  dp2 = dp1
Row40:  End If
Row41:  Range("A3").Value = sc
Row42:  Range("A4").Value = dp1
Row43:  Next
Row44:  Call Restore(R, T)
Row45:  Range("A5").Value = dp2
Row46:    i = i + 1
Row47:  Loop Until (dp2 = dp)
Row48:  Main = 1
Row49:  End Function
```

Function loads the data in the same way as discussed previously; chrom is initialized with attribute indexes. Note that we have taken TotalCClasses and SetDClasses in to QuickReduct method instead of "CalculateDRR". This measure is taken to enhance the performance as dependency will be calculated again and again, so this will avoid the declaration of the array again and again. However "ClrTCC" will be called every time dependency will be calculated to clear the previously created equivalence class structure.

In Row27, we have calculated dependency using the entire set of conditional attributes. Row31 copies the array "R" into "T". "R" represents reducts obtained so far. Row32 copies the attributes {Chrom – R} into {X}, i.e. all the attributes of Chrom that are not in R will be copied to X. The loop at Row33 iterates through all the attributes in "X". Row35 clears the previous equivalence class structure. Row36 first combines the current attribute represented by "tmp" array with "R" and then passes this to calculateDRR function which actually calculates dependency that is stored in "dp1". Now if dp1 is greater than dp2 which represents the previous highest dependency (it means that combining current attribute with R has increased the degree of dependency), then the current attribute is combined (concatenated) with R. dp2 is then updated by dp1. After the loop completes, we have "R" combined with attribute that provides the highest degree of dependency increase. So we copy "T" (which contains highest degree attribute concatenated with R) into R. The process continues until the dp2 becomes equal to "dp", the dependency of "D" on the entire set of conditional attributes.

8.9.1 *Miscellaneous Methods*

"CUD" method combines (concatenates) two integer arrays. It performs "union" operation. The following is the listing of the method (Listing 8.23).

Listing 8.23: CUD Method

```
Function CUD(C() As Integer, D() As Integer) As Integer()
Dim csize As Integer
Dim dsize As Integer
Dim cd() As Integer
If (isArrayEmpty(C)) Then
   Call Restore(cd, D)
   CUD = cd
   Exit Function
End If
If (isArrayEmpty(D)) Then
   Call Restorc(cd, C)
   CUD = cd
   Exit Function
End If
ReDim cd(1 To UBound(C) + UBound(D))
csize = UBound(C) - LBound(C) + 1
dsize = UBound(D) - LBound(D) + 1
Dim X As Integer
For X = LBound(C) To UBound(C)
cd(X) = C(X)
Next
```

(continued)

Listing 8.23 (continued)

```
Row22:  Dim j As Integer
Row23:  X = X – 1
Row24:  For j = 1 To UBound(D)
Row25:     cd(X + j) = D(j)
Row26:  Next
Row27:  CUD = cd
Row28:  End Function
```

Function takes two arguments i.e. two integer arrays and concatenates them. First it checks if array "C" is empty then copies array "D" into "cd". If "D" is empty then "C" is copied into "cd". If none of these cases exist, then we define "cd" to be equal the size of "C" plus size of "D" and then copies both arrays in "cd".

8.9.2 Restore Method

It restores source array "S" into target array "T". The following is the listing of the function (Listing 8.24).

Listing 8.24: Restore Method

```
Row1:   Function Restore(ByRef T() As Integer, ByRef S() As Integer) As
        Integer
Row2:   If (isArrayEmpty(S)) Then
Row3:   Restore = 0
Row4:   Exit Function
Row5:   End If
Row6:   ReDim T(1 To UBound(S))
Row7:   Dim i As Integer
Row8:   i = 1
Row9:   While (i < = UBound(S))
Row10:     T(i) = S(i)
Row11:     i = i + 1
Row12:  Wend
Row13:  Restore = 1
Row14:  End Function
```

Function first checks whether the source array is empty in which case this function returns zero. Otherwise, it defines the target array to be equal to the size of source array "S" and copies it to target array element by element.

8.9.3 *C_R Method*

This method copies one array into another by skipping the mentioned elements. The following is the listing of this function (Listing 8.25).

Listing 8.25: C_R Method

```
Public Function C_R(tmp() As Integer, C() As Integer, R() As Integer) As Integer
If (isArrayEmpty(R)) Then
   Call Restore(tmp, C)
   C_R = 0
   Exit Function
End If
ReDim tmp(1 To UBound(C) - UBound(R))
Dim Tmpi, Ci, Ri As Integer
Dim found As Boolean
Tmpi = 1
For Ci = 1 To UBound(C)
found = False
   For Ri = 1 To UBound(R)
If (C(Ci) = R(Ri)) Then
   found = True
End If
   Next
If (found = False) Then
      tmp(Tmpi) = C(Ci)
      Tmpi = Tmpi +1
End If
Next
C_R = 1
End Function
```

It takes three arguments: "tmp", the reference of the array in which final result will be copied, "C" array which will be copied and "R" array whose elements (also present in "C") will be skipped.

If "R" is empty, then the entire of "C" will be copied to "tmp" using the "Restore" function; otherwise tmp is defined equal to the size of "C" minus the size of "R". The function then iterates through each element in "C"; the outer loop controls this. The currently selected element is then searched in "R" (by the inner loop); if element is found, it is skipped; otherwise it is copied to "tmp".

8.10 Summary

In this chapter we have presented, source code of the provided API library and its description. All the API functions are explained up to sufficient details to use them or modify with any of the feature selection or RST-based algorithm. Each function was explained with details of the tasks it provides and explanation of important statements and data structures used. For some complex data structure, diagrammatic description was also provided.

Printed in the United States
By Bookmasters